Jordi Recasens

Indistinguishability Operators

Studies in Fuzziness and Soft Computing, Volume 260

Editor-in-Chief

Prof. Janusz Kacprzyk
Systems Research Institute
Polish Academy of Sciences
ul. Newelska 6
01-447 Warsaw
Poland
E-mail: kacprzyk@ibspan.waw.pl

Further volumes of this series can be found on our homepage: springer.com

Vol. 244. Xiaodong Liu and Witold Pedrycz
Axiomatic Fuzzy Set Theory and Its Applications, 2009
ISBN 978-3-642-00401-8

Vol. 245. Xuzhu Wang, Da Ruan, Etienne E. Kerre
Mathematics of Fuzziness – Basic Issues, 2009
ISBN 978-3-540-78310-7

Vol. 246. Piedad Brox, Iluminada Castillo, Santiago Sánchez Solano
Fuzzy Logic-Based Algorithms for Video De-Interlacing, 2010
ISBN 978-3-642-10694-1

Vol. 247. Michael Glykas
Fuzzy Cognitive Maps, 2010
ISBN 978-3-642-03219-6

Vol. 248. Bing-Yuan Cao
Optimal Models and Methods with Fuzzy Quantities, 2010
ISBN 978-3-642-10710-8

Vol. 249. Bernadette Bouchon-Meunier, Luis Magdalena, Manuel Ojeda-Aciego, José-Luis Verdegay, Ronald R. Yager (Eds.)
Foundations of Reasoning under Uncertainty, 2010
ISBN 978-3-642-10726-9

Vol. 250. Xiaoxia Huang
Portfolio Analysis, 2010
ISBN 978-3-642-11213-3

Vol. 251. George A. Anastassiou
Fuzzy Mathematics: Approximation Theory, 2010
ISBN 978-3-642-11219-5

Vol. 252. Cengiz Kahraman, Mesut Yavuz (Eds.)
Production Engineering and Management under Fuzziness, 2010
ISBN 978-3-642-12051-0

Vol. 253. Badredine Arfi
Linguistic Fuzzy Logic Methods in Social Sciences, 2010
ISBN 978-3-642-13342-8

Vol. 254. Weldon A. Lodwick, Janusz Kacprzyk (Eds.)
Fuzzy Optimization, 2010
ISBN 978-3-642-13934-5

Vol. 255. Zongmin Ma, Li Yan (Eds.)
Soft Computing in XML Data Management, 2010
ISBN 978-3-642-14009-9

Vol. 256. Robert Jeansoulin, Odile Papini, Henri Prade, and Steven Schockaert (Eds.)
Methods for Handling Imperfect Spatial Information, 2010
ISBN 978-3-642-14754-8

Vol. 257. Salvatore Greco, Ricardo Alberto Marques Pereira, Massimo Squillante, Ronald R. Yager, and Janusz Kacprzyk (Eds.)
*Preferences and Decisions,*2010
ISBN 978-3-642-15975-6

Vol. 258. Jorge Casillas and Francisco José Martínez López
Marketing Intelligent Systems Using Soft Computing, 2010
ISBN 978-3-642-15605-2

Vol. 259. Alexander Gegov
Fuzzy Networks for Complex Systems, 2010
ISBN 978-3-642-15599-4

Vol. 260. Jordi Recasens
Indistinguishability Operators, 2010
ISBN 978-3-642-16221-3

Jordi Recasens

Indistinguishability Operators

Modelling Fuzzy Equalities and Fuzzy Equivalence Relations

Author

Dr. Jordi Recasens
ETS Arquitectura del Vallès
Department of Mathematics and Computer Sciences
C. Pere Serra 1-15
08190 Sant Cugat del Vallès
Spain
E-mail: j.recasens@upc.edu

ISBN 978-3-642-42333-8 ISBN 978-3-642-16222-0 (eBook)

DOI 10.1007/978-3-642-16222-0

Studies in Fuzziness and Soft Computing ISSN 1434-9922

Typeset & Cover Design: Scientific Publishing Services Pvt. Ltd., Chennai, India.

Printed on acid-free paper

9 8 7 6 5 4 3 2 1

springer.com

To my wife María
To my daughter Clara

Foreword

In 1982, at a conference on Logic in Barcelona, I introduced the concept of a T-Indistinguishability operator in fuzzy logic. As far as I know, this was the first paper dealing with a concept I borrowed from the complementary views of Henri Poincaré's thinking on the physical continuum's intransitivity, Karl Menger's Indistinguishability Relations in Statistical Metric Spaces, and both Lotfi A. Zadeh and Enrique H.Ruspini's papers on Fuzzy Similarity Relations. Two years later, and with my then student Llorenç Valverde, we published in a book an essay on these operators in which we showed several conceptual examples, included their characterization by means of elemental fuzzy T-preorders, and posed the problem we called of Poincaré-Menger. In this book, Sergei Ovchinnikov also published a very interesting characterization of prod-Indistinguishability operators by means of positive 'measurements'.

I think this was the beginning of a subject which was then brilliantly continued by other people, including Joan Jacas, Dionís Boixader, and Jordi Recasens in Barcelona, starting with the 1987 Ph.D Dissertation of Joan Jacas, which is in an outstanding contribution. I cordially thank Jordi Recasens for conferring me the honor of writing the Foreword to his book *Indistinguishability Operators. Modelling Fuzzy Equalities and Fuzzy Equivalence Relations.*

In what follows let me make a few remarks on what I consider to be most important for conceptually capturing the essential treatment of the special fuzzy relations this book considers.

I. The idea behind T-Indistinguishability operators is to give a definition of a graded equivalence between the elements $a, b, c, ..$ in a set X. To this end, the problem is to establish a minimal number of properties a fuzzy relation $E : X \times X \to [0,1]$ does satisfy in order to express that $E(a,b)$ is a 'degree up to which a is *indistinguishable* from, or *equivalent*, to b', in such a way that crisp equivalence relations could be obtained as a special case. The typical reflexive and symmetric properties are easily translated by $E(a,a) = 1$, and $E(a,b) = E(b,a)$, for all a, b in X, but crisp transitivity must be extended by means of a formula involving the three degrees

$E(a,b), E(b,c)$, and $E(a,c)$. Given a continuous t-norm T , the fuzzy relation E is said to be T-transitive if $T(E(a,b), E(b,c)) \leq E(a,c)$, for all a, b, c in X, a definition clearly generalizing crisp transitivity since $T(1,1) = 1$. Notice that the bounds $r \leq E(a,b)$, and $r \leq E(b,c), r > 0$, give the weaker bound $T(r,r) \leq E(a,c)$, with $T(r,r) \leq r$. Of course, if T has not zero divisors then $T(r,r)$ is actually a bound for the indistinguishability degree between a and c, but if T is in the Łukasiewicz family, it is required that r satisfies $0 < T(r,r)$ since, otherwise we obtain the non informative result $E(a,c) \in [0,1]$. That is, if $T = W_\varphi$, it should be $r > \varphi^{-1}(0.5)$ to have $E(a,c) \in [T(r,r),1] \nsubseteq [0,1]$. It is interesting to note that the crisp relations $E(r)$, $r > 0$, defined by '$(a,b) \in E(r)$ iff $r \leq E(a,b)$', always reflexive and symmetric, are only crisp equivalences if $T = \min$, in which case they facilitate an indexed tree of crisp partitions of X.

II. From a conceptual point of view, it is relevant that indistinguishability operators E are directly related to the 'negation' of *distinguishability operators*, or pseudo-distances. For instance,

- E is a min-Indistinguishability operator *iff* $d = 1 - E$ is a ultradistance
- E is a W_φ or $prod_\varphi$-Indistinguishability operator *iff* $d = N_\varphi \circ E$ is such that $\varphi \circ d = 1 - \varphi \circ E$ is a pseudo-distance,

with the strong negation $N\varphi = \varphi^{-1} \circ (1 - id) \circ \varphi$, where φ is an order-automorphism of the unit interval. In the pseudo-metric space $(X, \varphi \circ d)$, it holds that $r \leq E(a,b)$ *iff* $(\varphi \circ d)(a,b) \leq 1 - \varphi(r)$. That is, for each a in X it holds that $r \leq E(a,b)$ *iff* b is in the neighborhood of a with radius $1 - \varphi(r)$. Similarly, for each b in X, it is $r \leq E(a,b)$ *iff* a is in the neighborhood of b with radius $1 - \varphi(r)$.

III. Regardless if the connectives are functionally expressible, commutative, associative, dual, the negation is a strong one, etc., the only algebras of fuzzy sets $([0,1]^X, \cdot, +, ')$ that are lattices are those with $\cdot = \min$ and $+ = \max$, none of them being a ortholattice and less again a boolean algebra, but only De Morgan algebras if $'$ is a strong negation. This is an important difference with the case of crisp sets, to which it should be added the inexistence of an specification axiom simple as in the case of crisp predicates, since the meaning of the imprecise ones is not definable by necessary and sufficient conditions and is context and purpose dependent.

What is widely considered is the idea that the most salient feature differentiating fuzzy from crisp sets is the failure of the principles of Non-contradiction (NC), and Excluded-middle (EM), although this is not a intrinsic property of fuzzy sets but only a property of their algebras. For instance, in the case of the standard algebras of fuzzy sets $([0,1]^X, T, S, N)$:

- NC holds *iff* $T = W_\varphi$, $N \leq N_\varphi$, and S is any continuous t-conorm
- EM holds *iff* $S = W_\psi^*$, $N_\psi \leq N$, and T is any continuous t-norm,

imply that both principles do hold if and only if $T = W_\varphi$, $S = W^*_\psi$, and $N_\psi \leq N \leq N_\varphi$, and fail otherwise. Nevertheless, by interpreting the principles *à la* Aristotle:

- NC : $\mu \cdot \mu'$ is always self-contradictory (impossible, in Aristotles terms), that is, $\mu \cdot \mu' \leq (\mu \cdot \mu')'$,
- EM : $(\mu + \mu')'$ is always self-contradictory ($\mu + \mu'$ is always the case, in Aristotles terms), that is, $(\mu + \mu')' \leq ((\mu + \mu')')'$,

for all μ in $[0,1]^X$, both principles do hold in any algebra of fuzzy sets $([0,1]^X, \cdot, +, ')$.

Hence, if it is not clear enough that the most important difference between fuzzy and crisp sets is the principles' failure, it should be admitted that, a clear difference lies, nevertheless, in the partial distinguishability between a fuzzy set μ and its complement μ'. For instance, if $X = [0,1]$, $\mu(x) = x$, $\mu'(x) = 1 - x$, and we consider the W-Indistinguishability operator $E(x,y) = 1 - |x - y|$, it holds that $E(\mu(x), \mu'(x)) = 1 - |2x - 1|$, which has value 1 *iff* $x = 0.5$, and 0 *iff* $x = 0$ or $x = 1$. That is, $0 < E(\mu(x), \mu'(x)) < 1$ *iff* $x \notin \{0, 0.5, 1\}$, although it holds NC: $(\mu \cdot \mu')(x) = W(\mu(x), \mu'(x)) = 0$ and EM: $(\mu + \mu')(x) = W^*(\mu(x), \mu'(x)) = 1$, for all x in $[0,1]$. This idea of intrinsically relating fuzziness to the non-complete distinguishability between a fuzzy set and its complement, was proposed many years ago by Ronald R. Yager, followed by myself, Claudi Alsina and Llorenç Valverde, and it is yet waiting to be developed further by, perhaps, considering the antonym instead of the complement, since the second does not correspond to a linguistic term.

IV. For representing precise concepts, crisp equivalence relations are the correct mathematical tool for the comprehension of the bipartition given by a predicate in the universe in which it is used. In the case of the representation of imprecise words by means of fuzzy sets, T-Indistinguishability operators are a mathematical tool for capturing the kind of imperfect fuzzy classification generated by an imprecise predicate. A nice example of this, is that of min-Indistinguishability operators that give, as described above, a tree of indexed crisp partitions of the universe, and could help to pose finite fuzzy-probability problems like classical partitions help to pose finite probabilistic ones.

There is again another topic for which the operators Recasens takes into account could be of a great interest. This *à la* Wittgenstein topic is not only important, but challenging for Computing with Words, and concerns the degree of 'family resemblance' existing between a predicate and their migrates to different universes of discourse.

The author of this interesting book deserves a high recognition for his effort in clearly presenting the theory of fuzzy equivalences and collecting, in a single volume, that which is currently dispersed in many articles published either in edited books, conference proceedings, or journals. The book is well organized, with very clear mathematical reasoning, and I am sure it will help people interested in fuzzy logic to better understand the world of ideas related to imprecise partitionings.

May this book receive all the acknowledgments and success it deserves!

Enric Trillas
Emeritus Researcher
European Centre for Soft Computing
Mieres(Asturias), Spain

Preface

The ability to determine an equality is essential to every theory because it is equivalent to the problem of distinguishing the objects that the theory deals with. This need -and the importance of selecting between different possibilities -arises from the fact that equality allows us to classify in the context of the theory; and to classify is one of the most important processes of knowledge since it allows us to relate, organize, generalize, find general laws, etc. In fact, a scientific knowledge that does not need an equality capable of classifying the objects it studies is unconceivable. Borges illustrates this very clearly in the following paragraph of *Funes the Memorious (Ficciones).*

> Éste, no lo olvidemos, era casi incapaz de ideas generales, platónicas. No sólo le costaba comprender que el símbolo genérico perro abarcara tantos individuos dispares de diversos tamaños y diversa forma; le molestaba que el perro de las tres catorce (visto de perfil) tuviera el mismo nombre que el perro de las tres y cuarto (visto de frente)... Sospecho... que no era muy capaz de pensar. Pensar es olvidar diferencias, es generalizar, abstraer. En el abarrotado mundo de Funes no había sino detalles, casi inmediatos. [1]

According to the Identity of indiscernibles, or Leibniz's law, two or more objects or entities are identical (i.e. they are one and the same entity) if they have all of their properties in common. In the abscence of uncertainty, this principle establishes equivalence relations and, in particular, the first Common Notion of *Euclid's Elements* or transitivity:

> Things which equal the same thing also equal one another.

[1] He was, let us not forget, almost incapable of general, platonic ideas. It was not only difficult for him to understand that the generic term dog embraced so many unlike specimens of differing sizes and different forms; he was disturbed by the fact that a dog at three-fourteen (seen in profile) should have the same name as the dog at three-fifteen (seen from the front)... I suspect... that he was not very capable of thought. To think is to forget differences, to generalize, to abstract. In the overly replete world of Funes there were nothing but details, almost contiguous details.

In the case of imprecision or vagueness, properties are fulfilled up to a degree, and because certain individuals are more similar than others, a gradual notion of equality appears.

As an example of this phenomenon let us compare two objects of an art collection according to their beauty. Due to the subjectivity and gradation of the concept *beauty*, a crisp equality cannot be expected, but a graded soft equality can.

A further example occurs when the fulfillment of properties is not a matter of degree but, instead, there are limitations on the perception and measurement of those properties. Let us, for instance, consider the case of a particular tool that provides measurements with an error margin ϵ. It naturally defines the following approximate equality relation $\sim$:

$$x \sim y \Leftrightarrow |x - y| \leq \epsilon.$$

Two measurements become distinguishable only if their absolute difference is greater than the error threshold ϵ. The relation $\sim$ is not transitive, since we could have $|x-y| \leq \epsilon$, $|y-z| \leq \epsilon$ and not necessarily $|x-z| \leq \epsilon$; this leads to Poincaré's paradox, which asserts that, in the real world, *equal* really means *indistinguishable* and therefore transitivity cannot generally be expected.

In these situations, where crisp equivalence relations are not sufficiently flexible to cope with uncertainty, the softening of equivalence and equality relations has proven to be a useful tool, and it is in this context that indistinguishability operators appear.

I have been working with indistinguishability operators for more than twenty years, and I am still amazed by the richness of their structure. This is because they can be seen from many different points of view.

- They represent a special kind of fuzzy relations, and the theory of fuzzy relations can be applied to them.
- Their very metric behaviour allows us to adapt the theory of metric spaces and topology to their study.
- They can be analyzed from the point of view of mathematical logic, since they also generalize (fuzzify) logic equivalence or biimplication.
- There is a close relationship between min-indistinguishability operators and hierarchical trees, which relates the former with Taxonomy and Classification Theory.

This book examines various aspects of indistinguishability operators. It is intended to be useful to people involved in either theory and applications who are interested in systems dealing with uncertainty and soft computing in general.

Bath, Sant Cugat del Vallès, Jordi Recasens
May 2010

Acknowledgements

I would like to thank the Universitat Politècnica de Catalunya for the sabbatical year that has allowed me the writing of this book. I am most grateful to Professor Jonathan Lawry for kindly inviting me to spend this sabbatical at the Department of Engineering Mathematics, University of Bristol and for the inspiring discussions that we held on the relation between indistinguishability, measure and prototype theory. I am also grateful to the Spanish Ministerio de Educación for funding my stay under the scheme 'Estancias de movilidad de profesores e investigadores españoles en centros extranjeros' PR2009-0079.

I wish to express my thanks to Claudi Alsina, Jordi Casabò, Pere Cruells, Jaume Lluis García-Roig, Piedad Guijarro, Amadeu Monreal, Rosa Navarro and María Santos Tomás, all members of the Department of Mathematics and Computer Science at the ETSAB, for the enjoyable atmosphere you can always find in it.

I want to express as well my gratitude to my colleagues Dionís Boixader, Maria Congost and Jesús Salillas for their support and understanding.

Most of the chapters of this book have been shaped by discussions with numerous colleagues over the past twenty years. I thank all of them, and especially to my colleagues and friends Joan Jacas, who introduced me to the world o fuzzy logic and uncertainty, Dionís Boixader, with whom I have had many exciting discussions on the topic and Enric Hernández from whom I have learnt new aspects of tackling with uncertainty.

I am indebted to professor Enric Trillas for writing the foreword of the book. It is an honour to have the book started by one of the fathers of indistinguishability operators.

Finally I'd like to give special thanks to my friends David and Kathleen Dando for their generosity and help during my stay in England.

Contents

1 Introduction

The notion of equality is essential in any formal theory since it allows us to classify the objects it deals with.

Classifying is one of the most important processes in knowledge, representation and inference since it permits us to relate, construct, generalize, find general laws, etc. It is inconceivable that a scientific knowledge should be without an equality that allows us to classify the objects it studies.

As a first approach to the concept of equality we can take Leibnitz's Law of Identity [89]:

> Two objects are identical if and only if they have all their properties in common in a given universe of discourse.

Note the relativism of this law, since two objects can be equal in one universe and different in any another. But for a fixed universe consisting of a set of elements and properties, Leibniz's Identity Law naturally generates an equivalence relation on that universe.

In many real situations the objects do not necessarily satisfy (or not) a property categorically, but rather satisfy it at some level or degree (think for example of the property *to be rich*). In these cases, properties are fuzzy concepts and the Identity Law is similarly fuzzy. We can not talk about identical objects, but a certain degree of similarity must be introduced. In this way, the equality becomes to a fuzzy concept. Hence, a model of equality useful in different branches of knowledge must be based on the concept of indistinguishability, since in a theory two element are considered as equal if they are indistinguishable at a certain level. Classical (crisp) equivalence relations are too rigid to model this kind of equality and hence it is necessary to introduce fuzzy equivalence relations.

These considerations lead us to the following central definition [144], [135].

Definition 1.1. *Let X be a universe and T a t-norm. A T-indistinguishability operator E on X is fuzzy relation $E : X \times X \to [0,1]$ on X satisfying for all $x, y, z \in X$*

J. Recasens: Indistinguishability Operator, STUDFUZZ 260, pp. 1–12.
springerlink.com

1. $E(x, x) = 1$ *(Reflexivity)*
2. $E(x, y) = E(y, x)$ *(Symmetry)*
3. $T(E(x, y), E(y, z)) \leq E(x, z)$ *(T-Transitivity)*

E separates points if and only if $E(x, y) = 1$ *implies* $x = y$.

$E(x, y)$ is interpreted as the degree of indistinguishability (or similarity) between x and y.

T-indistinguishability operators are also called *fuzzy equivalence relations* and T-indistinguishability operators separating points *fuzzy equalities.*

The above three properties fuzzify those of a crisp equivalence relation. Reflexivity expresses the fact that every object is completely indistinguishable from itself. Symmetry says that the degree in which x is indistinguishable from y coincides with the degree in which y is indistinguishable from x. Transitivity deserves a special attention. As early as in 1901, H. Poincaré showed interest in this property [112]. He stated that in the physical word, *equal* actually means *indistinguishable*, since when we assert that two objects are equal, the only thing we can be sure of is that it is impossible to distinguish them. This consideration leads to the paradox that two objects A and B can be considered as equal, B can be equal (indistinguishable) to C but in turn A and B can be different (distinguishable); i.e.: the following situation can happen:

$A = B$ and $B = C$, but $A \neq C$.

So Poincaré denies full transitivity in the real word.

T-transitivity in T-indistinguishability operators tries to overcome this paradox by considering degrees of indistinguishability between objects. It gives a threshold to $E(x, z)$ given the values for $E(x, y)$ and $E(y, z)$. The intuitive argument is that it is not reasonable in many cases to have three objects with high degree of indistinguishability between x and y and between y and z, but with x and z very distinguishable.

In a more logical context, transitivity expresses that the following proposition is true.

If x is indistinguishable from y and y is indistinguishable from z, then x is indistinguishable from z.

In order to be useful in modelling different equalities, transitivity should be flexible. This can be achieved with the selection of a particular t-norm.

In this sense, it is worth recalling that when we use the Product t-norm, we obtain the so-called probabilistic relations introduced and studied by K. Menger in probabilistic metric spaces [95]. If we choose the Łukasiewicz t-norm, we obtain the relations called likeness introduced by E. Ruspini [122]. For the minimum t-norm we obtain similarity relations [144]. It should also be noticed that a crisp equivalence relation is a T-indistinguishability operator for any t-norm T.

Indistinguishability operators determine the granularity of a system in the same way as equivalence relations do. An equivalence relation on a set X

determines a partition of X consisting of their equivalence classes. Only subsets of X corresponding to unions of these equivalence classes can be observed taking the equivalence relation into account. These are the information granules generated by it on X. There are different possibilities to generalize this idea to indistinguishability operators. For a T-indistinguishability operator E on a set X, Zadeh defined the fuzzy equivalence classes associated with E as its columns (also called the aftersets of E) i.e. the family $(\mu_x)_{x\in X}$ of fuzzy subsets of X defined for all $y \in X$ by $\mu_x(y) = E(x,y)$ [144]. It is a very natural generalization, since $\mu_x(y)$ gives the degree in which x and y are related by E. In [78] a theoretical justification of this definition is provided (see also [29]). Nevertheless, the fuzzy equivalence classes of E are not enough to determine the granularity generated by E on X. The observable fuzzy subsets with respect to E are called extensional subsets; i.e., fuzzy subsets μ of X satisfying $T(\mu(x), E(x,y)) \leq \mu(y)$ for all $x, y \in X$. This definition fuzzifies the predicate

If $x \in \mu$ and $x \sim y$, then $y \in \mu$.

For this reason, extensionality is the most important property a fuzzy subset can fulfill with respect to an indistinguishability operator and many parts of this book are devoted to its different aspects. Fuzzy points are a special kind of granules generated by E. A fuzzy point is an extensional fuzzy subset μ satisfying $T(\mu(x), \mu(y)) \leq E(x,y)$, which fuzzifies the predicate

If $x, y \in \mu$, then $x \sim y$.

A normal fuzzy point (i.e. with an element x satisfying $\mu(x) = 1$) is exactly a fuzzy equivalence class and in the crisp case points coincide with equivalence classes.

A different way to tackle the problem of partitioning a set was proposed by Ruspini [121] [10]. In his definition, no indistinguishability operators are needed and a finite family $\mu_1, \mu_2, .., \mu_n$ of normal fuzzy subsets of X is a T-S-partition of X when a) $T(\mu_i(x), \mu_j(x)) = 0$ and b) $S(\mu_1(x), \mu_2(x), ..., \mu_n(x)) = 1$ for all $x \in X$ and where T and S are a t-norm and a t-conorm respectively. a) assures the empty intersection of the elements of the partition, while b) says that they are a coverage of X. There exists a relation between these partitions and indistinguishability operators as can be found in [129]. See also [80], [85] and Chapter 3 for the relation between T-S-partitions, fuzzy points and fuzzy equivalence relations.

A very natural way to give semantics to fuzzy sets is using indistinguishability operators. The degree of membership of an object x of a universe X to a concept described by a fuzzy subset μ can be viewed as the degree of indistinguishability between x and the prototypes of μ. If P is the set of prototypes of the concept described by μ, then $\mu(x) = \sup_{p\in P} E(x,p)$ where E is the indistinguishability operator defining the similarity between the objects of X. This approach goes back to [6] and was first developed by Ruspini [121]. See also [41] [44]. It is easy to see that in fact μ is the smallest extensional fuzzy subset containing the prototypes P.

An interesting feature of indistinguishability operators is their metric behaviour. To every crisp equivalence relation $\sim$ with E its characteristic function, a pseudodistance (in fact a pseudo ultrametric) $d = 1 - E$ can be associated and vice versa. In this way there is a duality between equivalence relations and discrete distances. This duality is maintained between indistinguishability operators and a kind of Generalized metrics [133] called S-metrics (see Chapter 4). The corresponding S-metric is a classical pseudodistance if and only if the t-conorm S is smaller than or equal to the Łukasiewicz t-conorm. Also E is a similarity (a min-indistinguishability operator) if and only if the associated S-metric is a pseudoultrametric.

A second situation in which the metric aspects of T-indistinguishability operators are apparent is in the case of continuous Archimedean t-norms. If t is an additive generator of the continuous Archimedean t-norm T and $t^{[-1]}$ its pseudo inverse (see the Appendix), then $t \circ E$ is a pseudodistance and reciprocally, from a pseudodistance d the T-indistinguishability operator $t^{[-1]} \circ d$ can be obtained.

These previous results are consequence of the fact that T-indistinguishability operators are a special kind of Generalized metrics. A generalized metric space (X, d) consists of a set X and a map $d : X \times X \to S$ where $(S, *, \leq)$ is an ordered semigroup satisfying for all $x, y, z \in X$

1. $d(x, x) = 0$.
2. $d(x, y) = d(y, x)$.
3. $d(x, z) \leq d(x, y) * d(y, z)$.

Examples of Generalized metric spaces are ordinary metric spaces with $(S, *, \leq) = (\mathbb{R}^+, +, \leq)$, probabilistic metric spaces ($(S, *, \leq) = (\Delta^+, *, \leq)$ where Δ^+ is the set of positive distribution functions and $*$ the convolution) [127] and boolean spaces where X is a boolean space and $(S, *, \leq) = (X, \triangle, \leq)$ where $\triangle$ is the symmetric difference. T-indistinguishability operators are also Generalized metrics. Indeed (X, E) is a Generalized metric space with $(S, *, \leq) = ([0, 1], T, \leq_T)$ where $\leq_T$ is the reverse of the natural ordering in the unit interval so that 0 and 1 are the greatest and the smallest element respectively. This captures the idea that two objects are indistinguishable when their (generalized) distance is small. It also explains the previous results since the relation between T-indistinguishability operators and S-metrics and between T-indistinguishability operators and distances for continuous Archimedean t-norms is simply an isomorphism of Generalized metric spaces (see Chapter 4).

The book is organized as follows.

Chapter 2 presents the most popular ways to generate a T-indistinguishability operator. The first approach is based on calculating the transitive closure of a proximity relation (i.e. a reflexive and symmetric fuzzy relation). The sup $-T$ product is used to generate the operator and a topological approach then provides a theoretical justification is given. The Representation Theorem is a very natural way to generate indistinguishability operators from a family of fuzzy subsets. The importance of this result will be made clear throughout the book. In particular, minimal families generating an indistinguishability operator, called a basis, are important, since when the cardinality of these families (called its dimension) is small, then the information contained in the relation can be represented in a small number of fuzzy subsets. These fuzzy subsets also describe degrees of matching with some features or prototypes and give semantic meaning to the relation. Decomposable indistinguishability operators are generated by a fuzzy subset and it is interesting to note that they are equivalent to Mamdani implications in fuzzy control. The last approach presented in this chapter to generate indistinguishability operators is by calculating a transitive opening of a proximity relation R (an indistinguishability operator maximal among those smaller than or equal to R). An algorithm to find some of them is provided.

Extensionality is the most important property a fuzzy subset can satisfy with respect to an indistinguishability operator E, because extensional fuzzy subsets are the only observable fuzzy subsets and determine the granularity derived from E in the same way as in the crisp case only subsets which are the union of equivalence classes of an equivalence relation are observable with respect to it. Chapter 3 is devoted to the study of the set H_E of extensional fuzzy subsets of a T-indistinguishability operator E. An interesting result is that H_E coincides with the set of generators of E in the sense of the Representation Theorem. Two operators ϕ_E and ψ_E between fuzzy subsets of the universe of discourse X are introduced that for a given fuzzy subset of X produce its upper and lower approximation by extensional ones. This relates them to fuzzy rough sets and fuzzy modal logic. ϕ_E and ψ_E are also closure and interior operators and the fuzzy topological structure they generate on H_E is analyzed and related to the crisp topology associated to E. A third operator Λ_E is introduced that characterizes the fuzzy points of E. Indistinguishability operators between fuzzy subsets are defined and studied using the duality principle.

Indistinguishability operators have an important metric property. This is because they are a special kind of Generalized Metric Spaces since the unit interval with a t-norm is an ordered semigroup where 0 and 1 are the greatest and the smallest elements respectively. In Chapter 4 this metric behaviour is analyzed. In particular, the maps that preserve indistinguishability operators are given. The relation between indistinguishability operators with respect to isomorphic t-norms is studied and the isometries between them is related to their generators in the sense of the Representation Theorem. The duality between indistinguishability operators and S-metrics is established. Also the relation between indistinguishability operators and distances is given via the additive generators of continuous Archimedean t-norms.

Among the different T-indistinguishability operators, those defined with respect to the minimum t-norm present a particular behaviour. This is because their α-cuts are equivalence relations and therefore generate partitions on the universe of discourse. In this way they generate indexed hierarchical trees and are very useful in Cluster Analysis and Taxonomy. Chapter 5 is devoted to min-indistinguishability operators and their relation with hierarchical trees. The calculation of the different types of these operators is a difficult combinatorial problem which will be investigated for universes of low cardinalities in this chapter. Also a very simple way of storing them is presented. Their relation between min-indistinguishability operators and ultrametrics is also analyzed.

An expression of the metric nature of indistinguishability operators is that they generated metric betweenness relations if the t-norm is continuous Archimedean (Chapter 6). One dimensional T-indistinguishability operators are characterized by the fact that they generated linear betweenness relations. The length of an indistinguishability operator defined in this chapter relates the betweenness relations with the dimension and the sup $-T$ product. Roughly speaking, the greater the dimension, the smaller the cardinality of the betweenness relation and the length. Decomposable indistinguishability operators generate a special kind of betweenness relations called radial relations in which there is a particular element between any other two elements. In the real line, given a fuzzy number an indistinguishability operator is obtained that generates a betweenness relation compatible with the natural ordering on $\mathbb{R}$. If the values of a one-dimensional T-indistinguishability operator are distorted by some noise, then the betweenness relation generated by this new relation will probably be not linear and even can be empty. This means that the definition of betweenness relation can not capture the possibility of a relation to be "almost" linear or -more generally speaking- is not capable of dealing with points being "more or less" between others. Fuzzy betweenness relation are introduced in this chapter in order to overcome this problem.

The Representation Theorem states that every T-indistinguishability operator E on a set X can be generated by a family of fuzzy subsets of X. This family is not unique and a minimal family generating E is called a *basis* and its cardinality the *dimension* of E. Chapter 7 solves the problem of finding the dimension and a basis of a T-indistinguishability operator when T is a continuous Archimedean or the minimum t-norm. The first case is based on a geometrical interpretation of H_E while for the minimum t-norm the problem is solved in a combinatorial way. A fuzzy relation can be an indistinguishability operator for different t-norms. In this chapter an algorithm to decide if a fuzzy relation is one dimensional for some continuous Archimedean t-norm is provided. A method based on Saaty's reciprocal matrices to obtain a one dimensional T-indistinguishability operator close to a given one for the Product t-norm is generated. This is interesting since due to noise or imprecision, the values of a one dimensional indistinguishability operator can be distorted and in this way we can recover its original values.

In many cases there is a need to aggregate indistinguishability operators. This can be the case if the elements of a universe are defined by a number of features and there is an indistinguishability operator defined on each domain. Chapter 8 provides different ways to do this. Basically if the t-norm is continuous Archimedean, the weighted quasi-arithmetic mean of indistinguishability operators is also an indistinguishability operator. In addition, some kinds of OWA operators when used to aggregate indistinguishability operators also produce such an operator. Furthermore, there is a need to aggregate a non-finite number of indistinguishability operators. This is the case for example if we want to compare fuzzy subsets. In this chapter a method of aggregating fuzzy subsets of the real line (fuzzy quantities) is proposed.

The transitive closure $\overline{R}$ of a proximity relation R is greater than or equal to R while its transitive openings are smaller than or equal to R. If we want to approximate a proximity by a T-indistinguishability operator, there are other operators between them that will be closer to R. Chapter 9 presents different ways to find them when the t-norm is continuous Archimedean. One is finding the optimal as a weighted quasi-arithmetic mean of $\overline{R}$ and a transitive opening. A second modifies the values of the transitive closure or a transitive opening by applying a homotecy to the operator. A third approach uses non-linear programming techniques. A method to find good approximations for the minimum t-norm is also provided.

Fuzzy functions are useful tools in approximate reasoning. Roughly speaking they are functions compatible with indistinguishability operators on their domain and co-domain. The presence of these indistinguishability operators generates a granulation and in some sense a fuzzy function maps granules to granules. Their properties and structure are studied in Chapter 10. The existence of maximal fuzzy functions is proved and the fuzzy functions for which the indistinguishability operator on the domain is the classical equality are analyzed.

Two approaches to approximate reasoning in the presence of indistinguishability operators are given in Chapter 11. The first, based on IF-THEN rules, models the idea that from the rule "If x is A, then y is B" and x' is A' and A' "close" to A, we must entail that y' is B' with B' "close" to B. The second approach gives a theoretical background to fuzzy control based on the concept of a fuzzy function, the idea being that every fuzzy rule identifies a fuzzy patch and a fuzzy function can be generated by them.

Fuzzy subgroups were introduced by Rosenfeld in 1971 and vague groups by Demirci in 1999. Vague groups are a pair $(G, \overline{\circ})$ where $\overline{\circ}$ is a vague operation on G, $\overline{\circ}(a, b, c)$ meaning that c is vaguely or approximately $a \circ b$. This vague operation is compatible with a T-indistinguishability operator defined on G and its properties are closely related to those of fuzzy functions. In Chapter 12, vague groups are introduced and related to fuzzy subgroups. Given a crisp group $(G, \circ)$ and a T-indistinguishability operator defined on G, a vague group $(G, \overline{\circ})$ can be defined by $\overline{\circ}(a, b, c) = E(a \circ b, c)$. In this way there is a bijection between the set of normal fuzzy subgroups of G, the set T-indistinguishability operators on G invariant under translations and vague groups of G generated by this kind of indistinguishability operators on G. Under this bijections, if

$(G, \overline{\circ})$ corresponds to the normal fuzzy subgroup μ of μ, we can interpret $(G, \overline{\circ})$ as G/μ. In particular, the results are applied to the real line $(\mathbb{R}, +)$ and indistinguishability operators on $\mathbb{R}$ invariant under translations and with their columns being fuzzy numbers are characterized. A vague group is associated to every symmetric triangular fuzzy number and the idea of an integer number being vaguely a multiple of another is introduced.

The final chapter differs from earlier chapters in that it deals with finitely valued t-norms. These t-norms, valued on a finite totally ordered set L are very interesting, since they allow us to compute directly with linguistic variables. If we have for example a linguistic variable with labels *very cold, cold, neither cold nor warm, warm, hot, very hot*, then with a t-norm valued on these labels we can for instance calculate $T(\textit{very hot}, \textit{cold})$. Finitely valued indistinguishability operators are introduced in this chapter and some of their properties are studied. In particular, thanks to two pseudoinverses of an additive generator of a finitely valued t-norm, a Representation Theorem can be established for them and a way to find basis of them based on the solution of Diophantine systems of inequalities is provided. The relationship to indistinguishability operators valued on the unit interval is investigated and from that a new way to find a basis of a $[0, 1]$ valued indistinguishability operator or of an indistinguishability operator close to it is given.

An appendix with the basic definitions and properties of continuous t-norms has been added at the end of the book.

There are an important number of topics and applications of indistinguishability operators that have not been treated or only sketched on this volume, since the focus is mainly on theory. A number of the different applications of T-indistinguishability operators are outlined below.

- **Normalizing possibility distributions**
  It is well known that a normalized fuzzy subset of a universe X generates a possibility distribution on it. However, in many situations we have to handle non-normalized fuzzy subsets that do not generate valid possibility distributions. In these cases, it is assumed that there is a lack of information or evidence which means that the mass assignment values associated with the distribution add up to a number y_1 smaller that 1. The lack of mass $1 - y_1$ is then usually assigned to the empty set. In order to obtain a possibility distribution from a given non-normalized fuzzy subset μ, we have to reallocate the missing mass $1 - y_1$. The manner in which $1 - y_1$ is redistributed characterizes a specific normalization of μ and a number of such methods have been suggested. In [87] an axiomatic approach has been proposed in order to decide when a normalization can be considered as valid. In [118] the requirement that the normalization of a given fuzzy subset μ be as similar as possible to μ with respect to the natural similarity E_T on the set of fuzzy subsets of X (T a given t-norm) is investigated. The greatest (i.e. less specific) normal fuzzy subset $\hat{\mu}$ satisfying this reasonable property is then taken as the normalization of μ with respect to E_T. In this way interesting normalizations are generated. For example, the

minimal normalization, consisting of shifting the values y_1 to 1, the maximal normalization, consisting of adding $1 - y_1$ to all non-zero values of μ and the cross entropy normalization $\hat{\mu} = \frac{\mu}{y_1}$ correspond to the minimum, Łukasiewicz and Product t-norms respectively. For ordinal sums of copies of the Łukasiewicz t-norm intermediate normalizations are obtained.

- **F-transforms**
In [111] [25] the problem of approximating fuzzy relations is handled with the use of F-transforms. Let us only formulate the problem. Fixing $\epsilon \in [0, 1]$ there are fuzzy binary relations $R_1, , R_2, ..., R_n$ defined on $X_1, X_2, ..., X_n$ and an unknown fuzzy relation $f : X_1 \times X_2 \times ... \times X_n \rightarrow [0, 1]$ for which only partial information is available. Namely, there is a subset C of $X_1 \times X_2 \times ... \times X_n$ and the images $f(x_1, x_2, ..., x_n)$ are known for exactly the elements of C. Putting $R = T(R_1, R_2, ..., R_n)$ We want to find $D \subseteq X_1 \times X_2 \times ... \times X_n$ and $\hat{f} \in \mathbb{M}^D_{\vee(\wedge)}$ such that $E_T(\hat{f}(x_1, x_2, ..., x_n), f(x_1, x_2, ..., x_n)) \geq \epsilon$, $\forall (x_1, x_2, ..., x_n) \in C$, where

$$\mathbb{M}^D_{\vee} = \{\phi_R(g) \mid g : X_1 \times X_2 \times ... \times X_n \rightarrow [0, 1]\}$$

and

$$\mathbb{M}^D_{\wedge} = \{\psi_R(g) \mid g : X_1 \times X_2 \times ... \times X_n \rightarrow [0, 1]\}.$$

- **Kernels as indistinguishability operators**
In [99] [132] kernels which are positive-definite functions are proved to be T-indistinguishability operators for continuous Archimedean t-norms with additive generators $t(x) = \arccos x$ and $t(x) = \sqrt{1 - x}$. In [132] these results are applied to combinatorial chemistry for the design of large libraries of compounds in order to find new compounds with drug properties.
- **Indistinguishability operators and CAGD**
IF-THEN rules can be applied to CAGD (Computer Aided Geometric Design) in a straightforward way [65]. Let us illustrate the idea for generating a functional curve. In a similar way parametric curves and surfaces can also be built. Let us consider a set of control points $\{x_0, x_1, ..., x_n\}$ with their images $\{y_0, y_1, ..., y_n\}$. If we want continuity in our curve f, it is natural to impose that if x is close to x_i, then $f(x)$ be close to y_i. This can be expressed by a set of fuzzy rules similar to that of Chapter 11. If the fuzzy subsets of the rules and the t-norm are $\mathcal{C}^k$, then a Takagi Sugeno fuzzy controller gives a $\mathcal{C}^k$ curve. This approach is especially useful for generating surfaces when the control points are not located on a grid.
- **Concept lattices**
A triple (X, Y, I) where X and Y are sets and $I \in [0, 1]^{X \times Y}$ is called a formal context. The elements of X are considered as objects and the ones of Y as properties, $I(x, y)$ means the degree in which the object x satisfies property y. For every fuzzy subset μ of X the subset $\mu \uparrow$ of Y is defined by $\mu \uparrow (y) = \inf_{x \in X} \overrightarrow{T}(\mu(x) | I(x, y))$ and for every fuzzy subset ν of Y the fuzzy subset $\nu \downarrow$ of X is defined by $\nu \downarrow (x) = \inf_{y \in Y} \overrightarrow{T}(\nu(y) | I(x, y))$.

$(\uparrow, \downarrow)$ is called the polarity generated by I and the set $\mathcal{F}(X, Y, I)$ of fixed points (μ, ν) of $(\uparrow, \downarrow)$ is a fuzzy concept lattice. For a fuzzy concept $(\mu, \nu) \in \mathcal{F}(X, Y, I)$, μ consists of the objects of X that satisfy the properties of ν and ν is the set of properties shared by all the objects of μ. There is a bijection between the polarities between X and Y and fuzzy Galois connections, and the polarities are compatible with the natural indistinguishability operators E_T on $[0,1]^X$ and $[0,1]^Y$ ([8]).

- **Fuzzy modal logic**
  The use of T-indistinguishability operators, and especially of the maps ϕ_E, ψ_E in fuzzy modal logic has been briefly exposed in Chapter 3.
- **Fuzzy rough sets**
  Also these two maps generate fuzzy rough sets as explained in Chapter 3.
- **Observational entropy**
  In the definition of Shannon's entropy of a random variable ($H(X) = -\sum_{x \in X} p(x) \log_2(p(x))$ it is assumed that the elements x of X are completely indistinguishable. If a T-indistinguishability operator E is defined on X, then we cannot be sure if we have observed the event x or an event indistinguishable from x at some extent. In these situations, the definition of entropy must take E into account. In [141], [55] this idea has been developed and a new concept, called observational entropy has been studied. The observation degree of an element x_j of X is defined by $\pi(x_j) = \sum_{x \in X} p(x_j) E(x, x_j)$ and the observational entropy of X ($HO(X)$) by $HO(X) = -\sum_{x \in X} p(x) \log_2 \pi(x)$. From this definition, in [55] the concepts of simultaneous observation degree, conditioned observational entropy and joint observational entropy are given and applied to the generation of fuzzy decision trees.
- **Fuzzy decision trees**
  Decision trees have become one of the most relevant paradigms of machine learning methods. The main reason for this widespreading success lies in their proved applicability to a broad range of problems, in addition to appealing features as the readability of the knowledge represented in the tree. Therefore, a lot of work has been carried out from Quinlan's TDID3 algorithm in order to extend their applicability. An important possibility to do this is providing decision tree induction capable of coping with other sources of uncertainty beyond the probabilistic type. The case when uncertainty arises as a consequence of having indistinguishability operators on the domains of the attributes used to describe the set of instances can be found in [57]. There, observational entropy is used in building an observational decision tree from a set of examples. The problem could be posed as follows: Let $At = \{A_1, ..., A_n, C\}$ be a set of nominal attributes (being the classes of C the classification we want to learn), with domains $D_i = \{v_{i_1}, ..., v_{i_{mi}}\}$ and $D_c = \{v_{c_1}, ..., v_{c_{mc}}\}$. Let $S \subseteq D_1 \times ... \times D_n \times D_c$ be the set of instances and for each attribute A, we consider a T-indistinguishability operator E_A and a probability distribution p_A defined on the domain of A. From that a fuzzy decision tree is generated, the main

idea consisting in using the notion of conditioned observational entropy as a basis for the definition of the attribute selection measure, which leads to a criterion of observational information gain maximization.

- **Indistinguishability operators between prototypes**
Let us suppose the existence of some prototypes $a_1, a_2, ..., a_n$. It is common to assume that prototypes are completely different and that every prototype is completely distinguishable from the other ones, but it is not difficult to think of situations in which this is not the case. Let us then in this cases assume the presence of an indistinguishability operator $\overline{E}$ between them. If we consider a set X of objects resembling the prototypes to some extent, then it seems reasonable to extend the relation $\overline{E}$ to the set X. The preceding situation can be modeled in this way: There are n fuzzy subsets of X denoting the resemblance of the elements of X to the prototypes and an indistinguishability operator $\overline{E}$ between these prototypes. In [56] some methods to define an indistinguishability operator E on X compatible with $\overline{E}$ can be found depending on the properties of the fuzzy subsets defined on X.
- **Defuzzification**
Defuzzification is an essential problem in fuzzy systems that is always solved in a heuristic way. The first step in a fuzzy system consist on the fuzzification of the data. Starting from a crisp set $A \subseteq X$ or an element $x \in X$, a fuzzy subset is generated that takes the fuzziness of the system into account. In the case when a T-indistinguishability operator is defined on X, $\phi_E(A)$ and $\psi_E(A)$ are good candidates for fuzzifying A, since they are its best upper and lower approximations by fuzzy extensional subsets. The problem of defuzzification can be thought as the inverse problem of fuzzification. From an output, given as a fuzzy subset ν a crisp set is searched that can better express it once the fuzziness is eliminated. In [19] a couple of methods to defuzzify a fuzzy output are provided. The first one assumes the presence of a T-indistinguishability operator F on the set of outputs Y and defines two possible defuzzifications of a fuzzy subset ν: $\overline{defuzz}(\nu) = \inf\{C \subseteq Y \mid \nu \leq \phi_F(C)\}$ and $\underline{defuzz}(\nu) = \sup\{C \subseteq Y \mid \phi_F(C \leq \nu\}$. The second one finds the greatest crisp set closer to ν by the natural indistinguishability operator E_T. It is worth noticing that the defuzzification of ν is a crisp set, rather than a crisp element. If an element is needed, then the problem of selecting a specific element of the set is outside of the fuzzy system.
- **Generating aggregation operators**
When we aggregate two values a and b we may want to get a number λ which is as similar to a as to b or, in other words, λ should be equivalent to both values. If a T-indistinguishability operator E is defined on our universe, then the aggregation λ of a and b should satisfy $E(a, \lambda) = E(b, \lambda)$. This idea is developed in [76] when the universe is the unit interval $[0, 1]$ and the T-indistinguishability operator is the natural T-indistinguishability operator E_T associated to a continuous t-norm T. It

is proved that for a continuous Archimedean t-norm T with additive generator t the aggregation operator associated to E_T is the quasi-arithmetic mean m generated by t ($m(x,y) = t^{-1}\left(\frac{t(x)+t(y)}{2}\right)$). This can give a justification for using a concrete quasi-arithmetic mean in a real problem, since it will be related to a logical system having T as conjunction (and E_T as bi-implication). When T is an ordinal sum, interesting aggregation operators are obtained since the way they aggregate two values varies locally: For points in a piece $[a_i, b_i]^2$ where we have a copy of a continuous Archimedean t-norm with additive generator t_i their aggregation is related to the quasi-arithmetic mean generated by t_i while points outside these pieces with the smallest coordinate c in some $[a_i, b_i]$ have $a_i + (b_i - a_i)\, t_i^{-1}\left(\frac{t(\frac{c-a_i}{b_i-a_i})}{2}\right)$ as aggregation. The aggregation of the rest of the points is their smallest coordinate. The idea is generalized to weighted aggregations and aggregations of more than two objects.

- **Fuzzy logic in the narrow sense**
  Some fuzzy predicate logics need an equality that models fuzzy equality in fuzzy systems. This can be done in the syntactic level introducing an equality predicate $\approx$ on the language satisfying reasonable axioms:

  - $(\forall x)(x \approx x)$ (Reflexivity)
  - $(\forall x, y)(x \approx y \to y \approx x)$ (Symmetry)
  - $(\forall x, y, z)((x \approx y \mathbin{\&} y \approx z) \to x \approx z)$ (Transitivity)
  - $(\forall x_1, ..., x_k, y_1, ..., y_k)((x_1 \approx y_1 \mathbin{\&} ... \mathbin{\&} x_k \approx y_k) \to F_i(x_1, ..., x_k) \approx F_i(y_1, ..., y_k))$ where F_i is of arity k. (Congruence).

  In the interpretations, $\approx$ correspond to T-indistinguishability operators. In this setting, the properties of T-indistinguishability operators can be defined and derived at the formal language level. See [54],[52],[104],[49], [9] for different approaches.

Reflexive and T-transitive fuzzy relations are called T-preorders or T-fuzzy preorders. T-preorders are very interesting fuzzy relations since they fuzzify the concept of partial preorder on a set X. They have been studied in depth (see [12]) and since T-indistinguishability operators can be seen as special T-preorders (the symmetric ones), many of the ideas and results of this book can be applied to preorders in a straightforward way, though the concepts can have different meaning. For example, if P is a T-preorder, the images of ϕ_P (cf. Definition 3.7) fuzzify the concepts of filter and ideal and are called fuzzy filters and fuzzy ideals.

For the sake of simplicity we have assumed in this book that all t-norms are continuous, though most of the results remain valid for more general structures like left-continuous t-norms or GL-monoids [23]. The last chapter is an exception since it deals with finitely valued t-norms and finitely valued T-indistinguishability operators.

2

Generating Indistinguishability Operators

One of the most interesting issues related to indistinguishability operators is their generation, which depends on the way in which the data are given and the use we want to make of them. The four most common ways are:

- By calculating the T-transitive closure of a reflexive and symmetric fuzzy relation (a proximity or tolerance relation).
- By using the Representation Theorem.
- By calculating a decomposable operator from a fuzzy subset.
- By obtaining a transitive opening of a proximity relation.

In many situations, data come packed as a reflexive and symmetric fuzzy matrix or relation R, also known as a proximity or tolerance relation . When, for coherence, transitivity is also required, the relation R must be replaced by a new relation R' that satisfies the transitivity property. The transitive closure of R is the smallest of such relations among those greater than or equal to R. It is the most popular approximation of R and there are several algorithms for calculating it. In Section 2.1, the sup $-T$ product is introduced to generate it. If a lower approximation is required, transitive openings are a possibility 2.4.

A T-indistinguishability operator E_μ can be generated in a very natural way from a fuzzy subset μ of a set X. The Representation Theorem states that all T-indistinguishability operators can be generated from a family of fuzzy subsets of X. This means that, from the degrees of satisfiability of a family of features by the elements of X or their degree of similarity to a family of prototypes, a T-indistinguishability operator can be obtained. Section 2.2 will examine the Representation Theorem and Chapter 7 will look at some important consequences of the Theorem.

Another way to obtain a T-indistinguishability operator from a fuzzy subset is by calculating its corresponding decomposable operator. This is interesting due to the fact that these relations are used in Approximate Reasoning and in Mamdani fuzzy controllers. Chapter 6 will examine the betweenness

J. Recasens: Indistinguishability Operator, STUDFUZZ 260, pp. 13–39.
springerlink.com

relations they generate and compare them with those obtained using the Representation Theorem.

2.1 Transitive Closure

Given a t-norm T, the transitive closure of a reflexive and symmetric fuzzy relation R on a set X is the smallest T-indistinguishability operator relation $\overline{R}$ on X greater than or equal to R. There are several algorithms to compute it for the finite case, but in this chapter we will only study the $\sup -T$ product.

In the crisp case, if R is a crisp reflexive and symmetric relation, its transitive closure is the smallest equivalence relations that contains R. If R is represented by a graph, its transitive closure is the smallest graph that contains R and with all its connected components complete subgraphs. This produces the well known chain effect or chaining: if $a_1, a_2, ..., a_n$ is a chain of vertexes such that every one is connected to the next one, then in its transitive closure a_1 is connected to a_n. Though this effect is non-desirable in general, since the meaning of the relation may be distorted by this effect, in some cases it is a usable tool as can be seen for instance in [130], Example 13.3 of [84].

In this section, the $\sup -T$ product will be used to generate the transitive closure of a reflexive and symmetric fuzzy relation. Some properties of this product will be given. In particular, the set of fuzzy relations on a set is proved to be an ordered topological semigroup.

The $\sup - \min$ product has theoretical justifications (from set theory or from graph theory for instance) that are difficult to be generalized to justify the $\sup -T$ product for arbitrary t-norms. At the end of this section a natural topological approach to the $\sup - \min$ product and its generalization to the $\sup -T$ product will be given. The $\sup -T$ product will be identified to closure operators in V_D spaces ([127]) and in this way a theoretical basis for this product will be provided.

Definition 2.1. *A fuzzy relation R on a set X is a map $R : X \times X \to [0, 1]$.*

Given two fuzzy relations R, S on a set X, we will say that $R \leq S$ if and only if $R(x, y) \leq S(x, y)$ for all $x, y \in X$.

Definition 2.2. *Let R and S be two fuzzy relations on X and T a t-norm. The $\sup -T$ product of R and S is the fuzzy relation $R \circ S$ on X defined for all $x, y \in X$ by*

$$(R \circ S)(x, y) = \sup_{z \in X} T(R(x, z), S(z, y)).$$

In order to give a geometric interpretation to the $\sup -T$ product, let us give an example in the crisp case. If R is a crisp reflexive and symmetric relation on a finite set X, then it can be represented as a graph with as many vertices

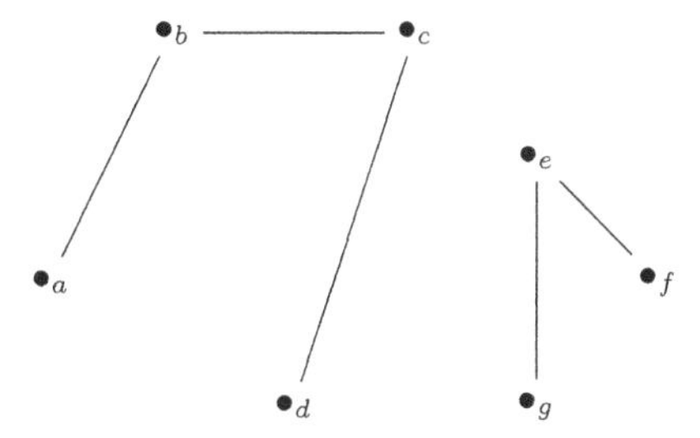

Fig. 2.1 A graph G_R corresponding to a crisp reflexive and symmetric relation R on $\{a, b, c, d, e, f, g\}$.

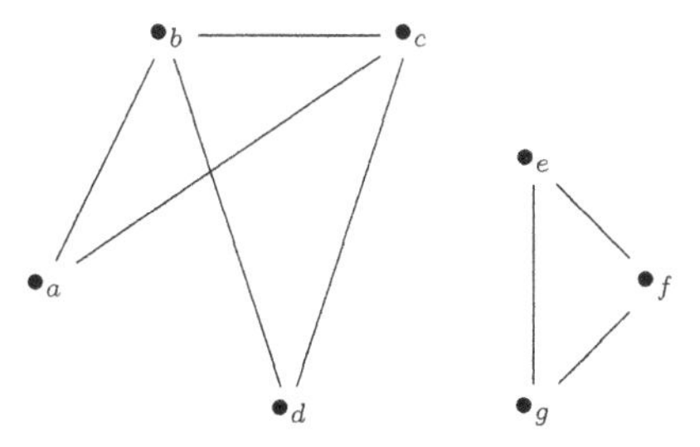

Fig. 2.2 The graph corresponding to $R \circ R$.

as the cardinality of X and where two vertices x and y are connected by an edge if and only if xRy. In this case $xR \circ Ry$ if and only if there exists z such that xRz and zRy. Figures 2.1 and 2.2 illustrate this case.

Since the $\sup -T$ product is associative for continuous t-norms, we can define for $n \in \mathbb{N}$ the nth power R^n of a fuzzy relation R:

$$R^n = \overbrace{R \circ \ldots \circ R}^{n\ times}.$$

Definition 2.3. *Let R_X be the set of reflexive and symmetric fuzzy relations on X. The distance d between $M, N \in R_X$ is defined by*

$$d(M, N) = \sup_{x,y \in X} |M(x, y) - N(x, y)|.$$

d is indeed a distance since the supremum of distances is a distance as well.

Proposition 2.4. *Let T be a continuous t-norm and R_X the set of reflexive and symmetric fuzzy relations on X. $(R_X, \sup -T)$ is a topological ordered semigroup.*

Proof

- Associativity.

$$\begin{aligned}(M \circ (N \circ P))(x,y) &= \sup_{z \in X} T(M(x,z), (N \circ P)(z,y)) \\ &= \sup_{z \in X} T(M(x,z), \sup_{t \in X} T(N(z,t), P(t,y))) \\ &= \sup_{z,t \in X} T(M(x,z), N(z,t), P(t,y)) \\ &= \sup_{t \in X} T(\sup_{z \in X} T(M(x,z), N(z,t)), P(t,y)) \\ &= ((M \circ N) \circ P)(x,y).\end{aligned}$$

- Neutral element. The identity relation $I(x,y) = \begin{cases} 1 \text{ if } x = y \\ 0 \text{ if } x \neq y. \end{cases}$
- Continuity. Since T is defined in a compact set, T is also uniformly continuous. Therefore,

$$\forall \epsilon > 0 \; \exists \delta > 0 \text{ such that } \forall m, n, a, b \in [0,1]$$

$$\max(|m-a|, |n-b|) < \delta \Rightarrow |T(m,n) - T(a,b)| < \epsilon. \quad (*)$$

We want to prove that given two relations $A, B \in R_X$,

$$\forall \epsilon > 0 \; \exists \delta > 0 \text{ such that } \forall M, N \in R_X$$

$$\max(d(M,A), d(N,B)) < \delta \Rightarrow d(M \circ N, A \circ B) < \epsilon.$$

Given $\epsilon > 0$, let us choose δ satisfying (*). Then

$$\begin{aligned}& d(M \circ N, A \circ B) \\ &= \sup_{x,y \in X} \left| \sup_{z \in X} T(M(x,z), N(z,y)) - \sup_{z \in X} T(A(x,z), B(z,y)) \right| \\ &\leq \sup_{x,y,z \in X} |T(M(x,z), N(z,y)) - T(A(x,z), B(z,y))| \leq \epsilon.\end{aligned}$$

- Monotonicity. $M \leq M'$, $N \leq N' \Rightarrow M \circ N \leq M' \circ N'$ is a consequence of the monotonicity of T.

Corollary 2.5. *For any $n \in \mathbb{N}$, the map $f : R_X \to R_X$ defined by $f(M) = M^n$ is continuous and non-decreasing.*

Proof. f is the composition of continuous and non-decreasing maps.

Of course, a similar result can be obtained for the set of all fuzzy relations on X with the $\sup -T$ product.

Proposition 2.6. *Let T and T' be two isomorphic continuous t-norms (i.e. there exists $f : [0,1] \to [0,1]$ such that $f \circ T = T(f \times f)$ cf. Definition 4.9). Then f induces an isomorphism $f : (R_X, \sup -T) \to (R_X, \sup -T')$.*

Proof. Given $R \in R_X$, let us define $f(R)$ by $(f(R))(x,y) = f(R(x,y))$ for all $x, y \in X$.

$$\begin{aligned}(f(R \circ_T S))(x,y) &= f((R \circ_T S)(x,y)) \\ &= f(\sup_{z \in X} T(R(x,z), S(z,y))) \\ &= \sup_{z \in X} f(T(R(x,z), S(z,y))) \\ &= \sup_{z \in X} T'(f(R(x,z)), f(S(z,y))) \\ &= \sup_{z \in X} T'((f(R))(x,z), (f(S))(z,y)) \\ &= ((f(R) \circ_{T'} f(S))(x,y).\end{aligned}$$

Before defining the transitive closure of a fuzzy relation, let us first recall what its α-cuts are and the relation between them and crisp relations.

Definition 2.7. *Let R be a reflexive and symmetric fuzzy relation on a set X and $\alpha \in [0,1]$. The α-cut of R is the crisp relation $\sim_\alpha$ on X defined by*

$$x \sim_\alpha y \text{ if and only if } R(x,y) \geq \alpha.$$

$\sim_\alpha$ is a reflexive and symmetric (crisp) relation and therefore generates a covering of X. We will identify the covering with the relation $\sim_\alpha$.

Lemma 2.8

1. *If $\alpha \geq \beta$, then the α-cut of R is a refinement of the β-cut.*
2. *The 0-cut of R is X.*
3. *If $R \geq S$, then the α-cut of S is a refinement of the α-cut of R.*

Proof. Trivial.

Definition 2.9. *Let R be a crisp reflexive and symmetric relation on a set X. The transitive closure of R is the smallest equivalence relation $\overline{R}$ on X that contains R.*

If X is finite, the transitive closure can be calculated by the single linkage method.

Definition 2.10. *Let R be a reflexive and symmetric crisp relation on a finite set X. $a, b \in X$ are related by single linkage (with respect to R) and we will denote it by $a\overline{R}b$ if and only if there exists a chain $a = x_0, x_1, ..., x_{r-1}, x_r = b$ of elements of X such that $x_{i-1} R x_i$ for all $i = 1, 2, ..., k$.*

Clearly $\overline{R}$ is the transitive closure of R.

If R is represented by a graph G_R, then its transitive closure is the graph $G_{\overline{R}}$ obtained by completing all connected subgraphs of G_R. Figures 2.3 and 2.4 show the graph G_R generated by a reflexive and symmetric relation R and its corresponding transitive closure $G_{\overline{R}}$.

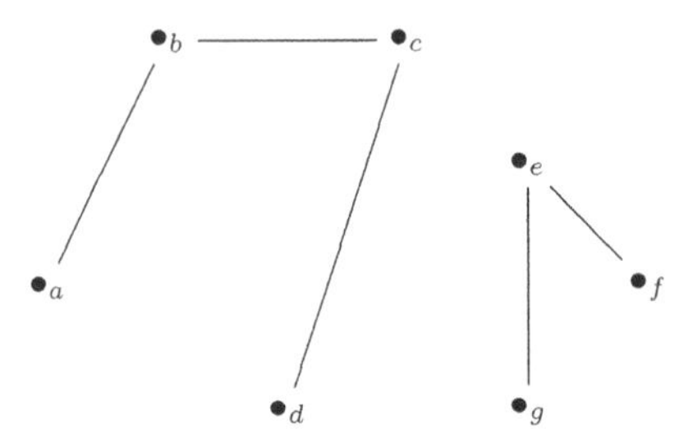

Fig. 2.3 A graph G_R corresponding to a crisp reflexive and symmetric relation on $\{a, b, c, d, e, f, g\}$.

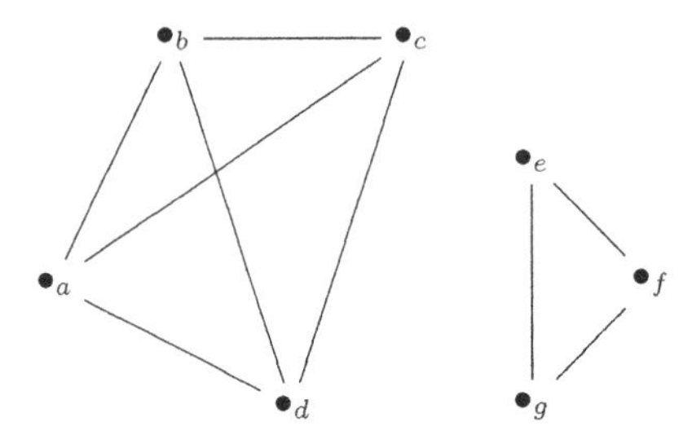

Fig. 2.4 The transitive closure $G_{\overline{R}}$ of G_R obtained by single linkage.

Definition 2.11. *Let T be a continuous t-norm and R a reflexive an symmetric fuzzy relation on a set X. Its transitive closure is the fuzzy relation $\overline{R}$ satisfying*

1. *$\overline{R}$ is a T-indistinguishability operator.*
2. *$R \leq \overline{R}$.*
3. *If S is a T-indistinguishability operator on X with $S \geq R$, then $S \geq \overline{R}$.*

The transitive closure of a reflexive and symmetric fuzzy relation R is the intersection of all T-indistinguishability operators greater than or equal to R.

Proposition 2.12. *Let T be a continuous t-norm and R a reflexive and symmetric fuzzy relation on X. Let A be the set of all T-indistinguishability operators on X greater than of equal to R. Then*

$$\overline{R} = \inf_{S \in A} S.$$

Proof. Let us first prove that $\inf_{S \in A} S$ is a T-indistinguishability operator.
Reflexivity and symmetry is trivial. Let us prove T-transitivity.

$$T(\inf_{S \in A} S(x, z), \inf_{S \in A} S(z, y)) \leq \inf_{S \in A} T(S(x, z), S(z, y)) \leq \inf_{S \in A} S(x, y).$$

Now, since $\overline{R} \leq S$ for all $S \in A$, then $\overline{R} \leq \inf_{S \in A} S$. But since $\overline{R} \in A$, then $\overline{R} \geq \inf_{S \in A} S$.

A very natural way to calculate the transitive closure of a fuzzy relation R is using the sup $-T$ product.

Lemma 2.13. *Let R be a reflexive and symmetric fuzzy relation on a set X and T a continuous t-norm. R is a T-indistinguishability operator if and only if $R \circ R \leq R$.*

Proof. R is T-transitive if and only if

$$T(R(x,z), R(z,y)) \leq R(x,y)$$

for all $x, y, z \in X$, which is equivalent to

$$(R \circ R)(x,y) = \sup_{z \in X} T(R(x,z), R(z,y)) \leq R(x,y).$$

Proposition 2.14. *Let R be a reflexive and symmetric fuzzy relation on a set X and T a continuous t-norm. Let R^T be the fuzzy relation on X defined by $R^T(x,y) = \sup_{n \in \mathbb{N}} R^n(x,y)$ for all $x, y \in X$. Then $R^T = \overline{R}$.*

Proof. Let us first prove that R^T is T-transitive.

$$\begin{aligned}(R^T \circ R^T)(x,y) &= \sup_{n \in \mathbb{N}} R^n(x,y) \circ \sup_{n \in \mathbb{N}} R^n(x,y) = \sup_{n \in \mathbb{N}, m \in \mathbb{N}} R^n(x,y) \circ R^m(x,y) \\ &= \sup_{n \in \mathbb{N}, m \in \mathbb{N}} R^{n+m}(x,y) \leq \sup_{k \in \mathbb{N}} R^k(x,y) = R^T(x,y).\end{aligned}$$

Now let S be a T-transitive fuzzy relation on X greater than or equal to R. By induction we will prove that $R^n \leq S$ for all $n \in \mathbb{N}$.

$$R^2 = R \circ R \leq S \circ S \leq S.$$

If $R^n \leq S$, then

$$R^{n+1} = R \circ R^n \leq S \circ S \leq S.$$

Hence

$$R^T(x,y) = \sup_{n \in \mathbb{N}} R^n(x,y) \leq S(x,y) \text{ for all } x, y \in X.$$

Lemma 2.15. *If R is a reflexive and symmetric fuzzy relation on X and T a continuous t-norm, then $R_T^n \leq R_T^m$ when $n \leq m$.*

Proof. Let us consider the classical equality I on X defined by

$$I(x,y) = \begin{cases} 1 \text{ if } x = y \\ 0 \text{ if } x \neq y \end{cases}$$

which is T-transitive. Since R is reflexive, $I \leq R$. Therefore $R^n = I \circ R^n \leq R \circ R^n = R^{n+1}$.

Proposition 2.16. *Let R be a reflexive an symmetric fuzzy relation on a finite set X of cardinality n. Then*

$$R^T = R_T^{n-1}.$$

Proof. If $x \neq y$, then

$$R^n(x,y) = \sup_{z_1, z_2, \ldots, z_{n-1}} T(R(x,z_1), R(z_1,z_2), ..., R(z_{n-1},y)).$$

Since the cardinality of X is n, at least two of the elements $x = z_0, z_1, z_2, ...,$ $z_{n-1}, z_n = y$ coincide. Let us suppose that $z_r = z_s$ with $r < s$. Then

$$\begin{aligned} &T(R(x_1,z_1), ..., R(z_{r-1},z_r), ..., R(z_s,z_{s+1}), ..., R(z_{n-1},y)) \\ &\leq T(R(x_1,z_1), ..., R(z_{r-1},z_r), R(z_s,z_{s+1}), ..., R(z_{n-1},y)) \\ &\leq R^{n-1}(x,y). \end{aligned}$$

Hence $R^n(x,y) \leq R^{n-1}(x,y)$ and therefore $R^n(x,y) = R^{n-1}(x,y)$.

In Chapter 4 the following proposition will be proved.

Proposition 2.17. *Let R be a reflexive and symmetric fuzzy relation on a set X. R is a* min*-indistinguishability operator if and only if for all $\alpha \in [0,1]$ the α-cut of R is a partition of X (i.e. $\sim_\alpha$ is an equivalence relation on X).*

Corollary 2.18. *Let R be a reflexive and symmetric fuzzy relation on X and $\alpha \in [0,1]$. The transitive closure of the α-cut of R coincides with the α-cut of the* min*-transitive closure of R.*

So the transitive closure is a way to get an upper approximation of a given reflexive and symmetric fuzzy relation by a T-indistinguishability operator.

If the cardinality n of X is finite, we can represent a fuzzy relation R on X by a square $n \times n$ matrix. The matrix is symmetric if and only if R is. R is reflexive if and only if the diagonal of the matrix consists of ones.

Example 2.19. Let R be the fuzzy relation given by the matrix

$$\begin{pmatrix} 1 & 0.9 & 0.3 & 0.4 \\ 0.9 & 1 & 0.5 & 0.4 \\ 0.3 & 0.5 & 1 & 0.9 \\ 0.4 & 0.4 & 0.9 & 1 \end{pmatrix}.$$

The transitive closure with respect to the t-norms of Łukasiewicz, Product and minimum respectively are

$$\begin{pmatrix} 1 & 0.9 & 0.4 & 0.4 \\ 0.9 & 1 & 0.5 & 0.4 \\ 0.4 & 0.5 & 1 & 0.9 \\ 0.4 & 0.4 & 0.9 & 1 \end{pmatrix}$$

$$\begin{pmatrix} 1 & 0.9 & 0.45 & 0.4 \\ 0.9 & 1 & 0.5 & 0.45 \\ 0.45 & 0.5 & 1 & 0.9 \\ 0.4 & 0.45 & 0.9 & 1 \end{pmatrix}$$

$$\begin{pmatrix} 1 & 0.9 & 0.5 & 0.5 \\ 0.9 & 1 & 0.5 & 0.5 \\ 0.5 & 0.5 & 1 & 0.9 \\ 0.5 & 0.5 & 0.9 & 1 \end{pmatrix}.$$

The sup $-$ min product and the min-transitive closure of a reflexive and symmetric fuzzy relation R have natural interpretations in the set and in the graph theory that are difficult to generalize to other t-norms. This section will end with a topological interpretation of the sup $-$ min product that will be generalized to more general sup $-T$ products.

Given a reflexive and symmetric fuzzy relation R on a set X and $p, q \in X$, $R(p,q)$ can be interpreted as the degree of proximity between p and q. Since *proximity* is a topological concept, it is natural to try to find a topological structure to X through R. This point of view will allow us to identify the sup $-T$ product with closure operators in certain V_D spaces [127].

Definition 2.20. *Let R be a reflexive and symmetric fuzzy relation on a set X. Given $\alpha \in [0,1]$ and $p \in X$, for every $h \in (0,1]$ $N_p^\alpha(h)$ is the neighbourhood of p given by*

$$N_p^\alpha(h) = \{q \in X \text{ such that } R(p,q) > \alpha - h\}.$$

Proposition 2.21. *If U, V are neighbourhoods of p, then there exists a neighbourhood W of p such that $W \subseteq U \cap V$.*

Proof. Trivial.

A set with a family of neighbourhoods satisfying the last proposition is called a V_D space ([127]).

Proposition 2.22. *The structure defined on X is a topology for all $\alpha \in [0,1]$ if and only if R is a* min*-indistinguishability operator on X.*

Proof. We must prove that if $q \in N_p^\alpha(h)$, then there exists $h' \in [0,1]$ such that $N_q^\alpha(h') \subseteq N_p^\alpha(h)$.

Let us take $h' = h$. If $x \in N_q^\alpha(h)$, then $R(x,p) \geq \min(R(p,q), R(q,x)) \geq min(\alpha - h, \alpha - h) = \alpha - h$. Therefore $x \in N_p^\alpha(h)$.

There is the notion of contiguity in V_D spaces.

Definition 2.23. *Given $A \subseteq X$, $p \in X$ is contiguous to A (at level α) if and only if there exists $q \in A$ such that $R(p,q) \geq \alpha$.*

The concept of contiguity in V_D spaces allow us to define a Čech closure operator C^α on the power set $P(X)$ of X.

Definition 2.24. *$C^\alpha : P(X) \to P(X)$ assigns to every subset A of X the set $C^\alpha(A)$ of contiguous elements to A (at level α).*

Proposition 2.25

1. *$C^\alpha(\emptyset) = \emptyset$.*
2. *$C^\alpha(A \cup B) = C^\alpha(A) \cup C^\alpha(B)$.*
3. *$A \subseteq C^\alpha(A)$.*
4. *$A \subseteq B \Rightarrow C^\alpha(A) \subseteq C^\alpha(B)$.*
5. *$C^\alpha(A) = \bigcup_{p\in A} C^\alpha\{p\}$.*
6. *$\alpha \geq \alpha' \Rightarrow C^\alpha(A) \subseteq C^{\alpha'}(A)$.*

Proof. Straightforward.

Definition 2.26. *The subsets $A \subseteq X$ such that $C^\alpha(A) = A$ are called C^α-closed.*

In order to define a topology on X we must define a Kuratowski closure operator C_K^α .

Definition 2.27. *$C_K^\alpha : P(X) \to P(X)$ assigns to every subset A of X the set $C_K^\alpha(A)$ intersection of all C^α-closed sets that contain A.*

Proposition 2.28

1. *R is a* min-*indistinguishability operator on X if and only if $C^\alpha = C_K^\alpha$ for all $\alpha \in [0,1]$.*
2. *$A \subseteq C^\alpha(A) \subseteq C_K^\alpha(A)$.*
3. *$(C^\alpha)^n(A) \subseteq C_k^\alpha(A)$ where $(C^\alpha)^n(A) = \overbrace{C^\alpha(C^\alpha(...(C^\alpha(A))))}^{n \text{ times}}$.*
4. *$C_K^\alpha(A) = \{x \in X$ such that $\exists n \in \mathbb{N}$ with $x \in (C^\alpha)^n(A)\}$.*

Proof

1. is a consequence of Proposition 2.22.
2. Trivial.
3. Trivial.
4. Let $M = \{x \in X$ such that $\exists n \in \mathbb{N}$ with $x \in (C^\alpha)^n(A)\}$.
   a. $M \subseteq C_K^\alpha(A)$: if $x \in M$, then there exists $n \in \mathbb{N}$ such that $x \in (C^\alpha)^n(A)$ and thanks to 3 $x \in C_K^\alpha(A)$.
   b. $C_K^\alpha(A) \subseteq M$: It is enough to see that M is a C^α-closed set, which is a consequence of 2.25.3.

From the last property the following four propositions follow.

Proposition 2.29. *If X is a finite set of cardinality s, then $(C^\alpha)^n = C_K^\alpha$ if $n \geq s$.*

Proposition 2.30. *Given* $p, q \in X$*, either* $C_K^\alpha(\{p\}) = C_K^\alpha(\{q\})$ *or* $C_K^\alpha(\{p\}) \cap C_K^\alpha(\{q\}) = \emptyset$*.*

Therefore C_K^α defines a partition on X. If $\sim_\alpha$ is the equivalence relation associated to this partition we have the following result.

Proposition 2.31. $\sim_\alpha$ *is the equivalence relation obtained by single linkage of the* α*-cut of* R*.*

Proposition 2.32. *Given* $n \in \mathbb{N}$ *and* $p, q \in X$*, if* $R^n_{\min}$ *is the* n*-th* $\sup - \min$ *power of* R*, then*

$$R^n_{\min}(p, q) = \inf \{\alpha \in [0, 1] \text{ such that } q \in (C^\alpha)^n(\{p\})\}.$$

These two last propositions, and especially Proposition 2.32, give a topological approach to the $\sup - \min$ product identifying it with Čech closure operators. This approach will be generalized now to general continuous t-norms.

Definition 2.33. *Given a reflexive and symmetric fuzzy relation* R *on a set* X*, a continuous t-norm* T *and* $\alpha \in [0, 1]$*, for any* $n \in \mathbb{N}$ $C_T^{\alpha,n} : X \to P(X)$ *is defined by*

$$C_T^{\alpha,n}(p) = \{q \in X \text{ such that } \exists\ x_0, x_1, ..., x_n \in X \text{ with } x_0 = p, x_n = q \text{ and } T(R(x_0, x_1), R(x_1, x_2), ..., R(x_{n-1}, x_n)) \geq \alpha\}.$$

In particular, $C_T^{\alpha,1}(p) = \{q \in X \text{ such that } R(p, q) \geq \alpha\}$.

$C_T^{\alpha,n}$ defined on X can be extended to $P(X)$ in the standard way.

Definition 2.34. $C_T^{\alpha,n} : P(X) \to P(X)$ *is defined by*

$$C_T^{\alpha,n}(A) = \bigcup_{p \in A} C_T^{\alpha,n}(p).$$

It is easy to prove that $C_T^{\alpha,n}$ satisfies the properties of Proposition 2.25. In particular, $C_T^{\alpha,n}$ is a Čech closure operator on the V_D space defined on X by the neighbourhoods $(N_p^\alpha)_{p \in X}$.

Also if T is the minimum t-norm, then $C_T^{\alpha,n} = (C^\alpha)^n$.

It is also easy to prove a similar result to Proposition 2.29.

Proposition 2.35. *If* X *is a finite set of cardinality* s*, then*

$$C_T^{\alpha,n} = C_T^{\alpha,n+1} \text{ if } n \geq s.$$

The next proposition relates the $\sup - T$ product to closure operators of a V_D space.

Proposition 2.36. *Let R be a reflexive and symmetric fuzzy relation on a set X. Given $n \in \mathbb{N}$ and $p, q \in X$,*

$$R_T^n(p,q) = \inf\{\alpha \in [0,1] \text{ such that } q \in C_T^{\alpha,n}(p)\}.$$

2.2 The Representation Theorem

The Representation Theorem allows us to generate a T-indistinguishability operator on a set X from a family of fuzzy subsets of X and, reciprocally, states that every T-indistinguishability operator can be obtained in this form.

Let us recall the concept of residuation of a t-norm. In fuzzy logic, the conjunction is usually modeled by a t-norm and its residuation is one of the possible ways to model the implication.

Definition 2.37. *The residuation $\overrightarrow{T}$ of a t-norm T is the map $\overrightarrow{T} : [0,1] \times [0,1] \to [0,1]$ defined for all $x, y \in [0,1]$ by*

$$\overrightarrow{T}(x|y) = \sup\{\alpha \in [0,1] \mid T(x,\alpha) \leq y\}.$$

The residuation of a t-norm T is also known as its quasi inverse, especially in early papers, and is also denoted by $\widehat{T}$.

The following properties of the residuation of a t-norm will be used throughout the book.

Lemma 2.38. *Given a left-continuous t-norm T, we have:*

1. *$\overrightarrow{T}(x|y)$ is left continuous and non increasing with respect to the first variable x.*
2. *$\overrightarrow{T}(x|y)$ is right continuous and non decreasing with respect to the second variable y.*

Proof. Trivial.

Lemma 2.39. *Given a left-continuous t-norm T, for any $x, y, z \in [0,1]$ the following relations hold:*

1. *$\overrightarrow{T}(1|x) = x$.*
2. *$x \leq y \Rightarrow \overrightarrow{T}(x|y) = 1$.*
3. *$T(x, \overrightarrow{T}(x|y)) \leq y$.*

Proof. Trivial.

Lemma 2.40. *Given a left-continuous t-norm T, for any $x, y, z \in [0,1]$ the following relation holds:*

$$\overrightarrow{T}(x|T(y,z)) \geq T(y, \overrightarrow{T}(x|z)).$$

Proof

$$\overrightarrow{T}(x|T(y,z)) = \sup\{\alpha|T(\alpha,x) \le T(y,z)\}.$$

From Lemma 2.39.4

$$T(y, \overrightarrow{T}(x|z), x) \le T(y,z).$$

Lemma 2.41. *Let T be a left continuous t-norm. Then*

$$\overrightarrow{T}(x|y) = \sup\{\alpha \in [0,1] \text{ such that } \overrightarrow{T}(\alpha|y) \ge x\}.$$

Proof. Let $A = \{\alpha \in [0,1]$ such that $T(\alpha,x) \le y\}$ and $B = \{\alpha \in [0,1]$ such that $\overrightarrow{T}(\alpha|y) \ge x\}$. Clearly $A \subseteq B$ and since T is left continuous, also $B \subseteq A$.

Lemma 2.42. *If T is a left continuous t-norm, then*

$$T(x, \overrightarrow{T}(y,z)) \le \overrightarrow{T}(\overrightarrow{T}(x|y)|z).$$

Proof

$$\begin{aligned} T(T(x, \overrightarrow{T}(y|z)), \overrightarrow{T}(x|y)) &= T(x, T(\overrightarrow{T}(x|y), \overrightarrow{T}(y|z))) \\ &\le T(x, \overrightarrow{T}(x|z)) \le z. \end{aligned}$$

Lemma 2.43. *Let T be a t-norm.*

$$\overrightarrow{T}(T(x,z)|T(y,z)) \ge \overrightarrow{T}(x|y).$$

Proof. Let us consider the sets

$$A_1 = \{\alpha \in [0,1] \text{ such that } T(\alpha,x) \le y\}$$

and

$$A_2 = \{\alpha \in [0,1] \text{ such that } T(\alpha, T(x,z)) \le T(y,z)\}.$$

If $\alpha \in A_1$, then $T(\alpha, T(x,z)) = T(T(\alpha,x),z)$ for all $z \in [0,1]$ and hence $\alpha \in A_2$.

So, $A_1 \subseteq A_2$ and

$$\overrightarrow{T}(x|y) = \sup A_1 \le \sup A_2 = \overrightarrow{T}(T(x,z)|T(y,z)).$$

Lemma 2.44. *Let T be a continuous Archimedean t-norm with additive generator t. Then*

$$\overrightarrow{T}(x|y) = t^{[-1]}(t(y) - t(x)) \; \forall x,y \in [0,1].$$

where $t^{[-1]}$ is the pseudo inverse of t (Theorem A.10).

Proposition 2.45. *The residuation $\overrightarrow{T}$ of a t-norm T is a T-preorder on $[0,1]$.*

Proof. It follows directly from the definition of residuation.

Corollary 2.46. *Let T be a left continuous t-norm. Then*

- $\overrightarrow{T}(\overrightarrow{T}(z|x)|\overrightarrow{T}(z|y)) \geq \overrightarrow{T}(x|y)$.
- $\overrightarrow{T}(\overrightarrow{T}(y|z)|\overrightarrow{T}(x|z)) \geq \overrightarrow{T}(x|y)$.

Proof. It is a consequence of the transitivity of $\overrightarrow{T}$.

The biimplication or logical equivalence is the symmetrized of the residuation.

Definition 2.47. *The biresiduation $\overleftrightarrow{T}$ of a t-norm T is the map $\overleftrightarrow{T}$: $[0,1] \times [0,1] \rightarrow [0,1]$ defined for all $x, y \in [0,1]$ by*

$$\overleftrightarrow{T}(x, y) = T(\overrightarrow{T}(x|y), \overrightarrow{T}(y|x)).$$

The biresiduation is also known as the natural T-indistinguishability operator associated to T and is also notated by E_T . This will be the notation used in this book in order to stress the fact that the biresiduation is a T-indistinguishability operator.

Lemma 2.48

a) $E_T(x, y) = \min(\overrightarrow{T}(x|y), \overrightarrow{T}(y|x))$.
b) $E_T(x, y) = \overrightarrow{T}(\max(x, y)| \min(x, y))$.

Proof

a) Either $\overrightarrow{T}(x|y) = 1$ or $\overrightarrow{T}(y|x) = 1$.
b)

$$\begin{aligned} E_T(x, y) &= \min(\overrightarrow{T}(\max(x, y)| \min(x, y)), \overrightarrow{T}(\min(x, y)| \max(x, y))) \\ &= \overrightarrow{T}(\max(x, y)| \min(x, y)). \end{aligned}$$

Example 2.49. See Table 2.1.

1. If T is a continuous Archimedean t-norm with additive generator t, then $E_T(x, y) = t^{-1}(|t(x) - t(y)|)$ for all $x, y \in [0,1]$.
   As special cases,
   - If T is the Łukasiewicz t-norm, then $E_T(x, y) = \overleftrightarrow{T}(x, y) = 1 - |x - y|$ for all $x, y \in [0,1]$.
   - If T is the Product t-norm, then $E_T(x, y) = \overleftrightarrow{T}(x, y) = \min(\frac{x}{y}, \frac{y}{x})$ for all $x, y \in [0,1]$ where $\frac{z}{0} = 1$.
2. If T is the minimum t-norm, then

$$E_T(x, y) = \overleftrightarrow{T}(x, y) = \begin{cases} \min(x, y) & \text{if } x \neq y \\ 1 & \text{otherwise.} \end{cases}$$

Proposition 2.50. *The natural T-indistinguishability operator with respect to a t-norm T is indeed a T-indistinguishability operator on $[0,1]$.*

Proof. Reflexivity and symmetry are trivial to prove.
Transitivity:

$$\begin{aligned} T(E_T(x,y), E_T(y,z)) &= T(T(\overrightarrow{T}(x|y), \overrightarrow{T}(y|x)), T(\overrightarrow{T}(y|z), \overrightarrow{T}(z|y))) \\ &\leq T(\overrightarrow{T}(x|z), \overrightarrow{T}(z|x)) = E_T(x,z). \end{aligned}$$

Table 2.1 The three most popular t-norms with their residuations and natural T-indistinguishability operators.

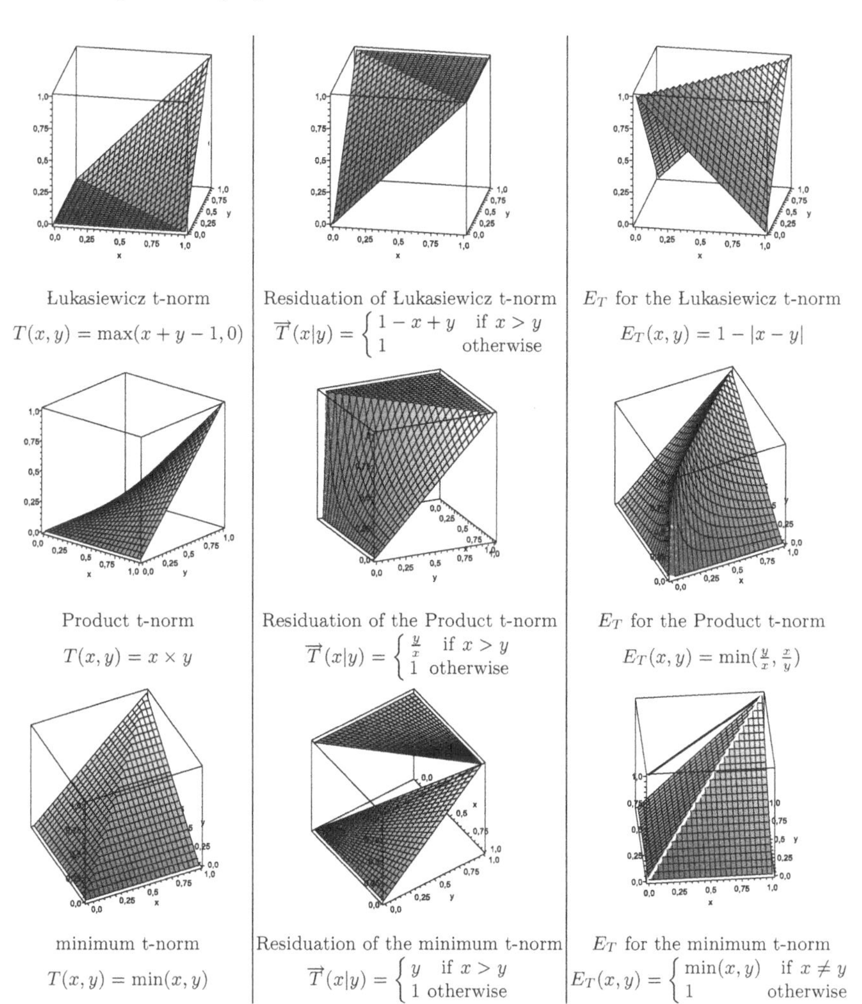

The following property will be used throughout the book.

Proposition 2.51. *Let E be a reflexive and symmetric fuzzy relation on a set X. E is a T-indistinguishability operator on X if and only if for all $x, y, z \in X$ $E_T(E(x,y), E(y,z)) \geq E(x,z)$.*

Proof

$$\begin{aligned} &E_T(E(x,y), E(y,z)) \\ &\quad = \min((\overrightarrow{T}(E(x,y)|E(y,z)), (\overrightarrow{T}(E(y,z)|E(x,y))) \geq E(x,z) \end{aligned}$$

if and only if

$$\overrightarrow{T}(E(x,y)|E(y,z)) \geq E(x,z) \text{ and } \overrightarrow{T}(E(y,z)|E(x,y)) \geq E(x,z)$$

which is equivalent to

$$T(E(x,z), E(x,y)) \leq E(y,z) \text{ and } T(E(x,z), E(y,z)) \leq E(x,y).$$

Every fuzzy subset μ of a set X generates a T-indistinguishability operator on X in a very natural way.

Lemma 2.52. *Let μ be a fuzzy subset of X and T a continuous t-norm. The fuzzy relation E_μ on X defined for all $x, y \in X$ by*

$$E_\mu(x,y) = E_T(\mu(x), \mu(y))$$

is a T-indistinguishability operator.

E_μ separates points if and only if μ is a one-to-one map.

Proof. E_μ is a T-indistinguishability operator since E_T is.

$$E_\mu(x,y) = 1 \text{ if and only if } E_T(\mu(x), \mu(y)) = 1 \text{ if and only if } \mu(x) = \mu(y).$$

Hence E_μ separates points if and only if μ is a one-to-one map.

In the crisp case, when $\mu = A$ is a crisp subset of X, E_A generates a partition of X into A and its complementary set $X - A$, since in this case $E_A(x,y) = 1$ if and only if x and y both belong to A or to $X - A$.

T-indistinguishability operators generated by a fuzzy subset as in the previous proposition are called one-dimensional.

Lemma 2.53. *Let $(E_i)_{i \in I}$ be a family of T-indistinguishability operators on a set X. The relation E on X defined for all $x, y \in X$ by*

$$E(x,y) = \inf_{i \in I} E_i(x,y)$$

is a T-indistinguishability operator.

Proof. Similar to Proposition 2.12.

The next theorem is a crucial result in order to understand the structure of T-indistinguishability operators. It will allow us to generate them from families of fuzzy subsets and reciprocally states that for any given T-indistinguishability operator families of fuzzy subsets generating it can be found.

Theorem 2.54. *Representation Theorem* [139] . *Let R be a fuzzy relation on a set X and T a continuous t-norm. R is a T-indistinguishability operator if and only if there exists a family $(\mu_i)_{i\in I}$ of fuzzy subsets of X such that for all $x, y \in X$*

$$R(x,y) = \inf_{i\in I} E_{\mu_i}(x,y).$$

$(\mu_i)_{i\in I}$ is called a generating family of R. A fuzzy subset belonging to a generating family of R is called a generator of R. In chapter 3 it will be proved that generators are exactly the extensional sets of R. In chapter 7 the generating families of R with minimal cardinality will be studied. These families are called basis of E and the cardinality of the corresponding set of indexes their dimension.

Proof. $\Leftarrow$)

Lemmas 2.52 and 2.53.

$\Rightarrow$)

For every $x \in X$ we can consider the fuzzy subset μ_x defined for all $y \in X$ by $\mu_x(y) = R(x,y)$. (This fuzzy subset is called the column or singleton of R associated to x). Then $R = \inf_{x\in X} E_{\mu_x}$.

Indeed,

$$E_{\mu_x}(y,z) = E_T(\mu_x(y), \mu_x(z)) = E_T(R(x,y), R(x,z)) \geq R(y,z).$$

So $E_{\mu_x} \geq R$ and $R \leq \inf_{x\in X} E_{\mu_x}$.

But for $y, z \in X$,

$$(\inf_{x\in X} E_{\mu_x})(y,z) \leq E_{\mu_y}(y,z) = E_T(R(y,y), R(y,z)) = R(y,z).$$

Hence $R = \inf_{x\in X} E_{\mu_x}$.

The Representation Theorem provides us with a method to generate a T-indistinguishability operator from a family of fuzzy subsets. These fuzzy subsets can measure the degrees in which different features are fulfilled by the elements of our universe X or can be the degree of compatibility with different prototypes.

Also, given a reflexive and symmetric fuzzy relation R on X, the T-indistinguishability operator $\underline{R}$ generated by the set of the columns of R can be built.

Proposition 2.55. $\underline{R} \leq R$.

Proof. For each $x, y \in X$, we have $E_{\mu_x}(x,y) = R(x,y)$. Therefore

$$\inf_{x \in X} E_{\mu_x} \leq R.$$

Example 2.56. Considering the same fuzzy relation R of Example 2.19 given by the matrix

$$\begin{pmatrix} 1 & 0.9 & 0.3 & 0.4 \\ 0.9 & 1 & 0.5 & 0.4 \\ 0.3 & 0.5 & 1 & 0.9 \\ 0.4 & 0.4 & 0.9 & 1 \end{pmatrix},$$

the obtained indistinguishability operators with respect to the t-norms of Łukasiewicz, Product and minimum respectively are

$$\begin{pmatrix} 1 & 0.8 & 0.3 & 0.4 \\ 0.8 & 1 & 0.4 & 0.4 \\ 0.3 & 0.4 & 1 & 0.9 \\ 0.4 & 0.4 & 0.9 & 1 \end{pmatrix}$$

$$\begin{pmatrix} 1 & 0.6 & 0.3 & 0.\widehat{3} \\ 0.6 & 1 & 0.\widehat{3} & 0.4 \\ 0.3 & 0.\widehat{3} & 1 & 0.75 \\ 0.\widehat{3} & 0.4 & 0.75 & 1 \end{pmatrix}$$

$$\begin{pmatrix} 1 & 0.3 & 0.3 & 0.3 \\ 0.3 & 1 & 0.3 & 0.4 \\ 0.3 & 0.3 & 1 & 0.3 \\ 0.3 & 0.4 & 0.3 & 1 \end{pmatrix}.$$

There is a similar Representation Theorem for fuzzy T-preorders that states

Theorem 2.57. *Representation Theorem for T-preorders [139]. Let R be a fuzzy relation on a set X and T a continuous t-norm. R is a T-preorder if and only if there exists a family $(\mu_i)_{i \in I}$ of fuzzy subsets of X such that for all $x, y \in X$*

$$R(x,y) = \inf_{i \in I} P_{\mu_i}(x,y)$$

where the fuzzy T-preorders P_{μ_i} on X are defined for all $x, y \in X$ by

$$P_{\mu_i}(x,y) = \overrightarrow{T}(\mu_i(x)|\mu_i(y)).$$

There is a generalization of the Representation Theorem due by Fodor and Roubens [45] that permits the generation of any T-transitive relation. This time two families of fuzzy subsets are needed.

Theorem 2.58. *Let R be a fuzzy relation on a set X and T a continuous t-norm. R is T-transitive if and only if there exist two families $(\mu_i)_{i\in I}$ and $(\nu_i)_{i\in I}$ of fuzzy subsets of X with $\mu_i \geq \nu_i$ $\forall i \in I$ such that for all $x, y \in X$*

$$R(x,y) = \inf_{i\in I} \overrightarrow{T}(\mu_i(x)|\nu_i(y)).$$

The Representation Theorems of this section are also valid for more general structures like left-continuous t-norms and GL-monoids [23].

2.3 Decomposable Indistinguishability Operators

Decomposable fuzzy relations have been applied successfully in Mamdani controllers, first using the minimum t-norm and then with more general t-norms. This section is focused on decomposable indistinguishability operators when the t-norm is continuous Archimedean or the minimum.

Definition 2.59. *For a given t-norm T, a fuzzy relation R on a set X is T-decomposable if and only if there exists a couple of fuzzy subsets μ, ν of X such that for all $x, y \in X$*

$$R(x,y) = T(\mu(x), \nu(y)).$$

We will say that the pair (μ, ν) generates R. If $\mu = \nu$, then we will simply say that μ generates R.

The first results will prove that a symmetric fuzzy relation can be generated by a single fuzzy subset.

The next lemma is trivial.

Lemma 2.60. *Let T be a t-norm, X a set and μ a fuzzy subset of X. The decomposable fuzzy relation R on X generated by μ is symmetric.*

Let us find which conditions a pair of fuzzy subsets must fulfill in order to generate a symmetric decomposable fuzzy relation.

Lemma 2.61. *Let T be a continuous Archimedean t-norm, t an additive generator of T, μ, ν two fuzzy subsets of a universe X and R the decomposable fuzzy relation on X generated by the pair (μ, ν). Let us assume that $R(x,y) \neq 0$ for all $x, y \in X$. R is symmetric if and only if*

1. $\mu \leq \nu$ and $\overrightarrow{T}(\nu|\mu)$ is a constant k
or
2. $\nu \leq \mu$ and $\overrightarrow{T}(\mu|\nu)$ is a constant k.

Proof. $\Rightarrow$)

Given $x, y \in X$, since $R(x,y) \neq 0$ and R is symmetric,

$$t(\mu(x)) + t(\nu(y)) = t(\mu(y)) + t(\nu(x)).$$

- If $\mu(x) \leq \nu(x)$, then

$$0 \leq t(\mu(x)) - t(\nu(x)) = t(\mu(y)) - t(\nu(y))$$

and $\mu(y) \leq \nu(y)$. Since this is true for all $y \in X$, $\mu \leq \nu$. Also

$$\overrightarrow{T}(\nu(y)|\mu(y)) = t^{-1}(t(\mu(y)) - t(\nu(y)) = t^{-1}(t(\mu(x)) - t(\nu(x)))$$

and therefore $\overrightarrow{T}(\nu|\mu)$ is a constant and we are in case 1.
- If $\mu(x) \geq \nu(x)$, a similar reasoning leads to case 2.

$\Leftarrow)$
If $\mu \leq \nu$ and $\overrightarrow{T}(\nu|\mu) = k$, then for all $x \in X$

$$\overrightarrow{T}(\nu(x)|\mu(x)) = t^{-1}(t(\mu(x)) - t(\nu(x))) = k$$

and

$$t(\mu(x)) - t(\nu(x)) = t(k).$$

Given $x, y \in X$,

$$\begin{aligned} R(x,y) &= T(\mu(x), \nu(y)) = t^{[-1]}(t(\mu(x)) + t(\nu(y))) \\ &= t^{[-1]}(t(k) + t(\nu(x)) + t(\mu(y)) - t(k)) \\ &= t^{[-1]}(t(\mu(y)) + t(\nu(x))) \\ &= T(\mu(y), \nu(x)) = R(y,x). \end{aligned}$$

In a similar way we can prove the symmetry of R when $\nu \leq \mu$ and $\overrightarrow{T}(\mu|\nu) = k$.

Lemma 2.62. *Let T be a continuous Archimedean t-norm with additive generator t and $x \in [0,1]$. Then*

$$x_T^{(\frac{1}{2})} = t^{-1}(\frac{1}{2}t(x)).$$

where $x_T^{(\frac{1}{2})}$ is the square root of x with respect to T (cf. Definition 9.11).

Proof

- If $x \neq 0$, then
  - $T(z,z) = x \Leftrightarrow t^{-1}(t(z) + t(z)) = x \Leftrightarrow 2t(z) = t(x) \Leftrightarrow z = t^{-1}(\frac{1}{2}t(x))$.
- If $x = 0$, then
  - if T is strict, then $x_T^{(\frac{1}{2})} = 0$ and $t^{-1}(\frac{1}{2}t(0)) = t^{-1}(\infty) = 0$.
  - if T is non-strict, let z be the greatest value in $[0,1]$ with $T(z,z) = 0$.

$$t^{[-1]}(2t(z)) = 0,\ 2t(z) = t(0),\ t(z) = \frac{1}{2}t(0),\ z = t^{-1}(\frac{1}{2}t(0)).$$

Lemma 2.63. *Let T be a continuous Archimedean t-norm with additive generator t and R a T-decomposable fuzzy relation on X with $R(x,y) \neq 0$ for all $x,y \in X$. R is symmetric if and only if there exists a pair (μ,ν) of fuzzy subsets of X such that for all $x,y \in X$*

1. $\mu \leq \nu$ and $R(x,y) = T(\rho(x),\rho(y))$ where $\rho(x) = T(\nu(x), \overrightarrow{T}(\nu(x)|\mu(x))^{\frac{1}{2}})$
or
2. $\nu \leq \mu$ and $R(x,y) = T(\rho(x),\rho(y))$ where $\rho(x) = T(\mu(x), \overrightarrow{T}(\mu(x)|\nu(x))^{\frac{1}{2}})$.

Proof. $\Rightarrow$)
Let (μ,ν) be a pair of fuzzy subsets of X generating R.

1. If $\mu \leq \nu$, then there exists a constant k such that $\overrightarrow{T}(\nu(x)|\mu(x)) = k$ and

$$t(\mu(x)) = t(\nu(x)) + t(k).$$

$$\begin{aligned}\overrightarrow{T}(\nu(x)|\mu(x))^{\frac{1}{2}} &= t^{-1}(\frac{1}{2}t \circ t^{-1}(t(\mu(x)) - t(\nu(x))) \\ &= t^{-1}(\frac{1}{2}(t(\mu(x)) - t(\nu(x)))\end{aligned}$$

and hence

$$\begin{aligned}\rho(x) &= T(\nu(x), \overrightarrow{T}(\nu(x)|\mu(x))^{\frac{1}{2}}) \\ &= t^{-1}(t(\nu(x)) + t \circ t^{-1}(\frac{1}{2}(t(\mu(x)) - t(\nu(x)))) \\ &= t^{-1}(t(\nu(x)) + \frac{1}{2}(t(\mu(x)) - t(\nu(x))) \\ &= t^{-1}\left(\frac{t(\nu(x)) + t\mu(x))}{2}\right) = t^{-1}\left(\frac{2t(\nu(x)) + t(k)}{2}\right).\end{aligned}$$

Therefore

$$\begin{aligned}R(x,y) &= t^{-1}(t(\mu(x)) + t(\nu(y))) \\ &= t^{-1}(t(\nu(x)) + t(k) + t(\nu(y))) \\ &= t^{-1}\left(t(\nu(x)) + \frac{t(k)}{2} + t(\nu(y)) + \frac{t(k)}{2}\right) \\ &= t^{-1}(t(\rho(x)) + t(\rho(y))) = T(\rho(x),\rho(y)).\end{aligned}$$

2. If $\nu \leq \mu$, a similar reasoning can be applied.

$\Leftarrow$)
Trivial. Simply take $\mu = \nu = \rho$.

ρ is is fact the quasi-arithmetic mean of μ and ν.

Definition 2.64. *m is a quasi-arithmetic mean in [0,1] if and only if there exists a continuous monotonic map $t : [0,1] \to [-\infty, \infty]$ such that for all $n \in \mathbb{N}$ and $x_1, ..., x_n \in [0,1]$*

$$m(x_1, ...x_n) = t^{-1}\left(\frac{t(x_1) + ... + t(x_n)}{n}\right).$$

m is continuous if and only if Ran $t \neq [-\infty, \infty]$.

Proposition 2.65. *With the same notations as in the previous lemma,*

$$\rho(x) = t^{-1}\left(\frac{t(\mu(x)) + t(\nu(x))}{2}\right) = m(\mu(x), \nu(x)).$$

Corollary 2.66. *Let T be a continuous Archimedean t-norm with additive generator t. If (μ, ν) and (μ', ν') generate the same symmetric T-decomposable fuzzy relation, then*

$$t^{-1}\left(\frac{t(\mu(x)) + t(\nu(x))}{2}\right) = t^{-1}\left(\frac{t(\mu'(x)) + t(\nu'(x))}{2}\right).$$

This means that the quasi-arithmetic mean of μ, ν and μ', ν' coincide.

So, symmetric decomposable fuzzy relations with respect to Archimedean t-norms can be generated by a single fuzzy subset ρ.

Corollary 2.67. *Let T be a continuous Archimedean t-norm. If R is a symmetric T-decomposable fuzzy relation on X with $R(x, y) \neq 0\ \forall x, y \in X$, then there is a fuzzy subset μ of X such that (μ, μ) generates X.*

Let us study symmetric decomposable fuzzy relations with respect to the minimum t-norm.

For two fuzzy subsets μ, ν of X let us define $m = \sup_{x \in X}\{\mu(x)$ such that $\mu(x) = \nu(x)\}$ if this set is not empty, and $m = 0$ otherwise.

Proposition 2.68. *Let R be the* min*-decomposable fuzzy relation on X generated by (μ, ν). R is symmetric if and only if*

1. $\mu \leq \nu$ and in the points $x \in X$ in which $\mu(x) \neq \nu(x)$, μ is a constant greater than or equal to m.
or
2. $\nu \leq \mu$ and in the points $x \in X$ in which $\nu(x) \neq \mu(x)$, ν is a constant greater than or equal to m.

Proof. $\Rightarrow$)

a) Let us first prove that either $\mu \leq \nu$ or $\nu \leq \mu$.
If there were $x, y \in X$ with $\mu(x) > \nu(x)$ and $\mu(y) < \nu(y)$, then the equality $\min(\mu(x), \nu(y)) = \min(\mu(y), \nu(x))$ would contradict the following two cases

- If $\mu(x) \geq \mu(y)$, then $\min(\mu(x), \nu(y)) = \nu(y)$. But from $\nu(y) > \mu(y)$, $\min(\mu(x), \nu(y)) > \min(\mu(y), \nu(x))$ would follow.
- If $\mu(x) \leq \mu(y)$, then $\min(\mu(x), \nu(y)) = \mu(x)$. But then we would have $\mu(x) = \nu(x)$.

Hence $\mu \leq \nu$ or $\nu \leq \mu$.

b) Let us examine both cases.

- Case $\mu \leq \nu$.
  If it does not exist $x \in X$ with $\mu(x) \neq \nu(x)$, then $\mu = \nu$ and the proof is ended.
  If there are $x, y \in X$ with $\mu(x) < \nu(x)$ and $\mu(y) < \nu(y)$, then $\min(\mu(x), \nu(y)) \leq \mu(x) < \nu(x)$ and $\min(\mu(y), \nu(x)) = \mu(y)$. Symmetrically, $\min(\mu(x), \nu(y)) = \mu(x)$ and $\mu(x) = \mu(y)$. So μ is constant in the points $x \in X$ where $\mu(x) \neq \nu(x)$.
  If there exists $z \in X$ with $\mu(z) = \nu(z)$, let $x \in X$ be a point with $\mu(x) < \nu(x)$.
  From $\min(\mu(x), \nu(z)) = \min(\mu(z), \nu(x))$, $\min(\mu(z), \nu(x)) = \nu(z)$ follows and therefore $\mu(x) \geq m$.
- Case $\nu \leq \mu$ is similar to the previous one.

$\Leftarrow$)

Case 1

- if $\mu(x) = \nu(x)$ and $\mu(y) = \nu(y)$, then trivially $\min(\mu(x), \nu(y)) = \min(\mu(y), \nu(x))$.
- if $\mu(x) = \nu(x)$ and $\mu(y) \neq \nu(y)$, then $\mu(x) = \nu(x) \leq \mu(y) < \nu(y)$ and therefore $\min(\mu(x), \nu(y)) = \min(\mu(y), \nu(x))$.
- if $\mu(x) \neq \nu(x)$ and $\mu(y) \neq \nu(y)$, then $\mu(x) = \mu(y) \leq \nu(x)$ and $\mu(x) = \mu(y) \leq \nu(y)$. Therefore $\min(\mu(x), \nu(y)) = \min(\mu(y), \nu(x))$.

Case 2 is similar to the previous one.

Now let us focus our attention on decomposable indistinguishability operators.

Proposition 2.69. *Let T be a t-norm and R a T-decomposable fuzzy relation on X. Then R is T-transitive.*

Proof. Let (μ, ν) be a pair of fuzzy subsets of X generating R. For $x, y, z \in X$,

$$\begin{aligned} T(R(x, y), R(x, z)) &= T(T(\mu(x), \nu(y)), T(\mu(y), \nu(z))) \\ &\leq T(T(\mu(x), 1), T(\nu(z), 1)) \\ &= T(\mu(x), \nu(z)) = R(x, z). \end{aligned}$$

Corollary 2.70. *Given a t-norm T, a decomposable fuzzy relation on X generated by a fuzzy subset μ of X is symmetric and T-transitive.*

Of course, a decomposable fuzzy relation generated by μ is reflexive if and only if $\mu \equiv 1$. Nevertheless, there is a possibility to generate a decomposable T-indistinguishability operator from any fuzzy subset μ.

Definition 2.71. *Let T be a t-norm. The decomposable T-indistinguishability operator E^μ generated by a fuzzy subset μ of X is defined for all $x, y \in X$ by*

$$E^\mu(x,y) = \begin{cases} T(\mu(x), \mu(y)) & \text{if } x \neq y \\ 1 & \text{otherwise.} \end{cases}$$

It is trivial to prove that E^μ is indeed a T-indistinguishability operator.

Proposition 2.72. *The decomposable T-indistinguishability operator on X generated by μ separates points if and only if there are no two different elements $x, y \in X$ with $\mu(x) = \mu(y) = 1$.*

Proof

$$E^\mu(x,y) = 1 \Leftrightarrow T(\mu(x), \mu(y)) = 1 \Leftrightarrow \mu(x) = \mu(y) = 1.$$

Decomposable indistinguishability operators generate interesting betweenness relations as it will be exposed in Chapter 6. They also generate another kind of relations that we call tetrahedric which characterize them.

Proposition 2.73. *Let T be a continuous Archimedean t-norm with additive generator t and μ a fuzzy subset of X. If the decomposable T-indistinguishability operator E^μ on X generated by μ satisfies $E^\mu(x,y) \neq 0$ for all $x, y \in X$, then it generates the following tetrahedric relation on X: Given four different elements $x, y, z, t \in X$,*

$$T(E^\mu(x,y), E^\mu(z,t)) = T(E^\mu(x,z), E^\mu(y,t)).$$

Proof

$$E^\mu(x,y) = T(\mu(x), \mu(y)) = t^{-1}(t(\mu(x)) + t(\mu(y)))$$

and

$$E^\mu(z,t) = T(\mu(z), \mu(t)) = t^{-1}(t(\mu(z)) + t(\mu(z))).$$

So,

$$\begin{aligned}
&T(E^\mu(x,y), E^\mu(z,t)) \\
&= t^{[-1]}(t \circ t^{-1}(t(\mu(x)) + t(\mu(y))) + t \circ t^{-1}(t(\mu(z)) + t(\mu(t)))) \\
&= t^{[-1]}(t(\mu(x)) + t(\mu(y)) + t(\mu(z)) + t(\mu(t))) \\
&= t^{[-1]}(t(\mu(x)) + t(\mu(z)) + t(\mu(y)) + t(\mu(t))) \\
&= t^{[-1]}(t \circ t^{-1}(t(\mu(x)) + t(\mu(z))) + t \circ t^{-1}(t(\mu(y)) + t(\mu(t)))) \\
&= T(E^\mu(x,z), E^\mu(y,t)).
\end{aligned}$$

The next proposition provides a partial reciprocal result.

Proposition 2.74. *Let T be a continuous Archimedean t-norm and E a T-indistinguishability operator separating points on X satisfying the tetrahedric relation and such that $E(x,y) > 0_T^{\frac{1}{2}}$ for all $x,y \in X$ and such that $\min_{x,y\in X} E(x,y)$ exists and is greater than 0. Then there exists a decomposable T-indistinguishability operator E' on X and a constant $k \in [0,1]$ such that*

$$E(x,y) = \overrightarrow{T}(k|E'(x,y)) \ \forall x,y \in X.$$

Proof. Let $a,b \in X$ $a \neq b$ be such that $E(a,b) = \min_{x,y\in X} E(x,y) > 0$.
For any two different $x,y \in X$ different from a and from b,

$$T(E(a,b),E(x,y)) = T(E(a,x),E(b,y)) = T(E(a,y),E(b,x))). \quad (*)$$

Let us consider the fuzzy subset μ of X defined by

$$\mu(x) = t^{-1}(\frac{t(E(a,x)) + t(E(b,x))}{2})$$

and the decomposable T-indistinguishability operators $E' = E^{\mu}$ it generates.
Then

$$E(x,y) = \overrightarrow{T}(E(a,b)|E'(x,y)).$$

Indeed, from (*)

$$E(x,y) = \overrightarrow{T}(E(a,b)|T(E(a,x),E(b,y))) = \overrightarrow{T}(E(a,b)|T(E(a,y),E(b,x))).$$

We must only show that

$$E'(x,y) = T(E(a,x),E(b,y)).$$

$$\begin{aligned}
&E'(x,y)\\
&= T(\mu(x),\mu(y))\\
&= t^{[-1]}(t(\mu(x)) + t(\mu(y)))\\
&= t^{[-1]}(t(t^{-1}(\frac{t(E(a,x)) + t(E(b,x))}{2})) + t(t^{-1}(\frac{t(E(a,y)) + t(E(b,y))}{2})))\\
&= t^{[-1]}(\frac{t(E(a,x)) + t(E(b,x))}{2} + \frac{t(E(a,y)) + t(E(b,y))}{2}).
\end{aligned}$$

On the other hand, thanks to (*),

$$\begin{aligned}
T(E(a,x),E(b,y)) &= t^{-1}(\frac{t(T(E(a,x),E(b,y))) + t(T(E(a,y),E(b,x)))}{2})\\
&= t^{-1}(\frac{t(E(a,x)) + t(E(b,y)) + t(E(a,y)) + t(E(b,x))}{2}).
\end{aligned}$$

2.4 Transitive Openings

As we have seen in section 2.1, the transitive closure of a reflexive and symmetric fuzzy relation R gives a T-indistinguishability operator greater than or equal to R. In this case it is possible to obtain the best upper approximation since the infimum of T-indistinguishability operators is also a T-indistinguishability operator. If we want a lower approximation, then the situation is more complicated since the supremum of indistinguishability operators is not such an operator in general. What we can find is T-indistinguishability operators maximal among the ones that are smaller than or equal to a given reflexive and symmetric fuzzy relation. These relations are not unique, but there can be an infinite quantity of them, even in sets of finite cardinality.

Unfortunately, there is no general method to calculate them. In [45], an algorithm to find maximal transitive openings of a given fuzzy relation is given, but the obtained openings are not symmetric in general and in the process of symmetrizing them, maximality can be lost. Heuristic methods to obtain T-indistinguishability operators smaller than or equal to a given fuzzy relation close to maximal ones have been proposed [31], but a general methodology to find them is still unknown.

The minimum t-norm is an exception because of the special behaviour of min-indistinguishability operators (see Chapter 5). In this case there are a number of algorithms to find at least some of the min-transitive openings of a given proximity relation. A classic method is the complete linkage that will be explained below. Other algorithms can be found in [26] [48]. In the complete linkage, the values of a proximity relation or matrix $R = (a_{ij})$ on a finite set X are modified according to the next algorithm to obtain a min-transitive opening. Given two disjoint subsets C_i C_j of X its similarity degree S is defined by $S(C_i, C_j) = \min_{i \in C_i, j \in C_j}(a_{ij})$. As usual in Cluster Analysis, the subsets of a partition of X will be called clusters. The complete linkage algorithm goes as follows.

1. Initially a cluster C_i is assigned to every element x_i of X (i.e. the clusters of the first partition are singletons).
2. In each new step two clusters are merged in the following way.
   If $\{C_1, C_2, ..., C_k\}$ is the actual partition, then we must select the two clusters C_i and C_j for which the similarity degree $S(C_i, C_j)$ is maximal. (If there are several such maximal pairs, one pair is picked at random). The new cluster $C_i \cup C_j$ replaces the two clusters C_i and C_j, and all entries of a_{mn} and a_{nm} of R with $m \in C_i$ and $n \in C_j$ are lowered to $S(C_i, C_j)$.
3. Step 2 is repeated until there remains one single cluster containing all the elements of X.

Example 2.75. Let us consider the proximity R on $X = \{x_1, x_2, x_3, x_4\}$ with matrix

$$\begin{array}{c} \\ x_1 \\ x_2 \\ x_3 \\ x_4 \end{array} \begin{array}{c} \begin{array}{cccc} x_1 & x_2 & x_3 & x_4 \end{array} \\ \begin{pmatrix} 1 & 0.1 & 0.7 & 0.4 \\ 0.1 & 1 & 0.4 & 0.3 \\ 0.7 & 0.4 & 1 & 0.5 \\ 0.4 & 0.3 & 0.5 & 1 \end{pmatrix} \end{array}.$$

The first partition is $C_1 = \{x_1\}$, $C_2 = \{x_2\}$, $C_3 = \{x_3\}$, $C_4 = \{x_4\}$. The greatest similarity degree between clusters is $S(C_1, C_3) = 0.7$. These two clusters are merged to form $C_{13} = \{x_1, x_3\}$. The matrix does not change in this step.

The new partition is C_{13}, C_2, C_4. The similarity degrees are

$$\begin{aligned} S(C_{13}, C_2) &= \min(a_{12}, a_{32}) = \min(0.1, 0.4) = 0.1 \\ S(C_{13}, C_4) &= \min(a_{14}, a_{34}) = \min(0.4, 0.5) = 0.4 \\ S(C_2, C_4) &= a_{24} = 0.3. \end{aligned}$$

The greatest similarity degree is 0.4 and the new partition is therefore $C_{134} = \{x_1, x_3, x_4\}$, $C_2 = \{x_2\}$. The entries a_{14}, a_{41}, a_{34}, a_{43} of the matrix R are replaced by 0.4 obtaining

$$\begin{array}{c} \\ x_1 \\ x_2 \\ x_3 \\ x_4 \end{array} \begin{array}{c} \begin{array}{cccc} x_1 & x_2 & x_3 & x_4 \end{array} \\ \begin{pmatrix} 1 & 0.1 & 0.7 & 0.4 \\ 0.1 & 1 & 0.4 & 0.3 \\ 0.7 & 0.4 & 1 & 0.4 \\ 0.4 & 0.3 & 0.4 & 1 \end{pmatrix} \end{array}.$$

In the last step, we merge the two clusters C_{134}, C_2. The similarity degree is

$$S(C_{134}, C_2) = \min(a_{12}, a_{32}, a_{42}) = \min(0.1, 0.4, 0.3) = 0.1.$$

The transitive opening of R obtained by complete linkage is then

$$\begin{array}{c} \\ x_1 \\ x_2 \\ x_3 \\ x_4 \end{array} \begin{array}{c} \begin{array}{cccc} x_1 & x_2 & x_3 & x_4 \end{array} \\ \begin{pmatrix} 1 & 0.1 & 0.7 & 0.4 \\ 0.1 & 1 & 0.1 & 0.1 \\ 0.7 & 0.1 & 1 & 0.4 \\ 0.4 & 0.1 & 0.4 & 1 \end{pmatrix} \end{array}.$$

3 Granularity and Extensional Sets

The presence of an indistinguishability operator on a universe determines its granules.

According to Zadeh, granularity is one of the basic concepts that underlie human cognition [146] and the elements within a granule 'have to be dealt with as a whole rather than individually' [145].

> Informally, granulation of an object A results in a collection of granules of A, with a granule being a clump of objects (or points) which are drawn together by indistinguishability, similarity, proximity or functionality [146].

In a classical (crisp) context, a crisp equivalence relation $\sim$ on a universe X determines the granules of X as its equivalence classes. Indeed, if we take the relation $\sim$ into account, only the sets that are unions of equivalence classes of $(X, \sim)$ can be observed in X, and properties are shared by all of the elements of the same equivalence class [18].

If the crisp equivalence relation is replaced by a fuzzy one, i.e. an indistinguishability operator E, there are basically two ways to consider the granularity induced on the universe by E.

- One is by considering the fuzzy equivalence classes generated by E, as defined by Zadeh in [144]. For each object x of the universe, the fuzzy equivalence class of x is the column μ_x of E ($\mu_x(\cdot) = E(x, \cdot)$). An extensive study of this idea can be found in [78].
- A more general extension of the granules is achieved by fuzzy points, the granularity consisting, then, of the fuzzy points of E.

A fuzzy point is an extensional fuzzy subset μ that satisfies $T(\mu(x), \mu(y)) \leq E(x, y)$, which fuzzifies the predicate

> If x and y belong to the same point μ, then they are equivalent with respect to the equivalence relation E.

J. Recasens: Indistinguishability Operator, STUDFUZZ 260, pp. 41–79.
springerlink.com

Fuzzy equivalence classes are exactly the normal fuzzy points of E and, in the crisp case, crisp equivalence classes are exactly the points associated with the crisp relation.

An important notion related to granularity and fuzzy equivalence relations is the idea of observability. In the crisp case again, if an equivalence relation is defined on a universe X, then only subsets of X that are unions of equivalence classes are compatible with the equivalence relation. Other subsets cannot be observed by taking the equivalence relation into account. In the fuzzy context, the observable fuzzy subsets with respect to a T-indistinguishability operator E are exactly the extensional fuzzy subsets of E.

From a structural point of view, it is especially interesting to study the set H_E of all extensional fuzzy subsets of a T-indistinguishability operator E defined on a set X. Section 3.1 is devoted to this and introduces two maps ϕ_E and ψ_E between fuzzy subsets of the universe of discourse. These maps are key tools for studying the structure of H_E because they characterize H_E as the set of their fixed points, and for a given fuzzy subset μ of X, $\phi_E(h)$ and $\psi_E(h)$ are the smallest extensional fuzzy subset greater than or equal to μ and the largest extensional fuzzy subset smaller than or equal to μ respectively and hence its upper and lower approximations in H_E [18]. H_E can be interpreted as the set of fuzzy subsets of the quotient set X/E (i.e.: $H_E = [0,1]^{X/E}$) and $\phi_E : [0,1]^X \to [0,1]^{X/E}$ can be interpreted as the canonical map. Note that if the indistinguishability operator E is a crisp one, then $\phi_{E|X}$ is the crisp canonical map $\pi : X \to X/E$.

ϕ_E and ψ_E, as upper and lower approximation operators of fuzzy sets by observable ones, can also be thought of as the key tools for defining fuzzy rough sets on X. They are also useful in other fields, such as fuzzy modal logic, where they model the possibility and necessity operators.

ϕ_E and ψ_E are a closure operator and an interior operator, respectively. Therefore, H_E can be seen as a fuzzy topology of X. Also, E generates a metric topology on X. Section 3.1 will analize the close relationship between these topologies.

A third map (Λ_E) is introduced in Section 3.2 in order to characterize the columns of E. The main results state that fuzzy points can be thought of as columns of extensions $(\overline{X}, \overline{E})$ of (X, E) and that the columns of E are the normal fixed points of Λ_E. The fixed points of Λ_E are the maximal fuzzy points of X. Also the set $\mathrm{Im}(\Lambda_E)$ is characterized as a set of fixed points of Λ_E^2.

Section 3.3 explores what happens when the elements of a family $(\mu_i)_{i\in I}$ of fuzzy subsets are columns of the T-indistinguishability operator that they generate using the Representation Theorem. It turns out that there is a close relationship between this property and fuzzy points.

In many applications, it is essential to have a way to measure the degree of similarity or indistinguishability between the fuzzy subsets of a universe. The most natural way to do this will set out in Section 3.4 using the duality principle. Other possibilities will be presented in Section 8.4.

3.1 The Set H_E of Extensional Fuzzy Subsets

Definition 3.1. *Let E be a T-indistinguishability operator on a set X. A fuzzy subset μ of X is extensional with respect to E (or simply extensional) if and only if for all $x, y \in X$*

$$T(E(x,y), \mu(y)) \leq \mu(x).$$

H_E will be the set of extensional fuzzy subsets of X with respect to E.

The previous definition fuzzifies the predicate

If x and y are equivalent and $y \in \mu$, then $x \in \mu$.

The set H_E has been widely studied [68], [69], [23] and its elements have been characterized as the generators [69] of E, the eigenvectors [68] i.e. the fixed points of ϕ_E and ψ_E, the logical states associated to E [4] and its extensional and observable sets [23].

If E is a crisp equivalence relation on X, then a crisp subset A of X is extensional if and only if it is the union (and intersections if we want to obtain the empty set) of the equivalence classes of A. H_E restricted to crisp subsets is in this case the set $\{0,1\}^{X/E}$ of subsets of the quotient set X/E.

Proposition 3.2. *Let E be a T-indistinguishability operator on X, μ a fuzzy subset of X and E_μ the T-indistinguishability operator generated by μ as in Lemma 2.52. $\mu \in H_E$ if and only if $E_\mu \geq E$.*

Proof. $E_\mu(x,y) = E_T(\mu(x), \mu(y)) \geq E(x,y)$ if and only if $T(E(x,y), \mu(x)) \leq \mu(y)$ and $T(E(x,y), \mu(y)) \leq \mu(x)$.

Hence H_E coincides with the set of generators of E.

Lemma 3.3. *Given a T-indistinguishability operator E on a set X and an element $x \in X$, the column μ_x of x is extensional.*

Proof. Given $y, z \in X$

$$E_{\mu_x}(y,z) = E_T(E(x,y), E(x,z)) \geq E(y,z).$$

Proposition 3.4. *Let E be a T-indistinguishability operator on a set X. The following properties are satisfied for all $\mu \in H_E$, $(\mu_i)_{i\in I}$ a family of extensional fuzzy subsets and $\alpha \in [0,1]$.*

1. $\bigvee_{i\in I} \mu_i \in H_E$.
2. $\bigwedge_{i\in I} \mu_i \in H_E$.
3. $T(\alpha, \mu) \in H_E$.
4. $\overrightarrow{T}(\mu|\alpha) \in H_E$.
5. $\overrightarrow{T}(\alpha|\mu) \in H_E$.

Proof

1. follows from the continuity of T.
2. follows from the continuity of T.
3. $T(E(x,y),T(\alpha,\mu)(y)) = T(\alpha,T(E(x,y),\mu(y)) \leq T(\alpha,\mu(x)) = T(\alpha,\mu)(x)$.
4. We must prove
$$T(E(x,y),\overrightarrow{T}(\mu(y)|\alpha)) \leq \overrightarrow{T}(\mu(x)|\alpha)$$
which is equivalent to prove
$$T(\mu(x),E(x,y),\overrightarrow{T}(\mu(y)|\alpha)) \leq \alpha.$$
But
$$T(\mu(x),E(x,y),\overrightarrow{T}(\mu(y)|\alpha)) \leq T(\mu(y),\overrightarrow{T}(\mu(y)|\alpha)) \leq \alpha.$$
5. We must prove
$$T(E(x,y),\overrightarrow{T}(\alpha|\mu(y))) \leq \overrightarrow{T}(\alpha|\mu(x))$$
which is equivalent to prove
$$T(\alpha,E(x,y),\overrightarrow{T}(\alpha|\mu(y))) \leq \mu(x).$$
But
$$T(\alpha,E(x,y),\overrightarrow{T}(\alpha|\mu(y)) \leq T(E(x,y),\mu(y)) \leq \mu(x).$$

The next Theorem 3.6 characterizes the sets of extensional sets with respect to T-indistinguishability operators as the sets satisfying the properties of the last proposition.

Lemma 3.5. *Let E be a T-indistinguishability operator on a set X. H_E is a generating family of E in the sense of the Representation Theorem 2.54.*

Proof. Trivial, since any generating family of E is contained in H_E.

Theorem 3.6. [23] *Let H be a subset of $[0,1]^X$ satisfying the properties of Proposition 3.4. Then there exists a T-indistinguishability operator E on X such that $H = H_E$. E is uniquely determined and it is generated (using the Representation Theorem) by the family of elements of H.*

Proof. Let E be the T-indistinguishability operator generated by H.

Let us first prove that $H = H_E$.

Since H is a generating family of E, its fuzzy subsets are extensional and $H \subseteq H_E$.

Given $\mu \in H_E$ and $y \in X$, let us define the fuzzy subset μ_y by $\mu_y(x) = T(E(x,y),\mu(y))$. We will show that $\mu = \sup_{y\in X} \mu_y$.

$$\begin{aligned}\mu_y(x) &= T(\inf_{\nu\in H} E_\nu(x,y),\mu(y))\\ &= \inf_{\nu\in H} T(\min(\overrightarrow{T}(\nu(x)|\nu(y)),\overrightarrow{T}(\nu(y)|\nu(x))),\mu(y)).\end{aligned}$$

Since $\mu(y)$ and $\nu(y)$ are constants, applying properties 2, 3, 4, 5 of 3.4 we have that $\mu_y \in H$.

Since μ is extensional, $\mu_y(x) = T(E(x,y),\mu(y)) \leq \mu(x)$. Adding that $\mu_y(y) = \mu(y)$ we obtain $\mu = \sup_{y\in X}\mu_y$ which is an element of H tanks to property 3.4.1. This gives that $H_E \subseteq H$.

Let us prove the uniqueness of E.

Let E' be a T-indistinguishability operator with $H = H_{E'}$. Then $E' \leq E$.

To prove the other inequality, let us consider for all $x \in X$ the column μ_x of E', which is extensional with respect to E' and therefore $\mu_x \in H$. But this means that μ_x is also extensional with respect to E and we have

$$E(x,y) = T(E(x,y),\mu_x(x)) \leq \mu_x(y) = E'(x,y).$$

Theorem 3.6 establish a bijection between T-indistinguishability operators and subsets H of $[0,1]^X$ satisfying the properties of Proposition 3.4. If E_H is the T-indistinguishability operator generated by H, then $E_{H_E} = E$ and $H_{E_H} = H$.

3.1.1 The Map ϕ_E

This and the next subsections introduce two maps ($\phi_E, \psi_E : [0,1]^X \to [0,1]^X$) which are key tools in order to study the structure of H_E [18].

The main result concerning these maps is that both, ϕ_E and ψ_E, have H_E as the set of fixed points.

Definition 3.7. *Let E be a T-indistinguishability operator on a set X. The map $\phi_E : [0,1]^X \to [0,1]^X$ is defined for all $x \in X$ by*

$$\phi_E(\mu)(x) = \sup_{y\in X} T\left(E(x,y),\mu(y)\right).$$

Proposition 3.8. *For all $\mu,\mu' \in [0,1]^X$,*

1. *If $\mu \leq \mu'$ then $\phi_E(\mu) \leq \phi_E(\mu')$.*
2. *$\mu \leq \phi_E(\mu)$.*
3. *$\phi_E(\mu \vee \mu') = \phi_E(\mu) \vee \phi_E(\mu')$.*
4. *$\phi_E(\phi_E(\mu)) = \phi_E(\mu)$.*
5. *$\phi_E(\{x\})(y) = \phi_E(\{y\})(x)$*
6. *$\phi_E(T(\alpha,\mu)) = T(\alpha,\phi_E(\mu))$.*

Proof

1. It is a consequence of the monotonicity of the t-norm.
2. $\phi_E(\mu)(x) = \sup_{y\in X} T\left(E(x,y),\mu(y)\right) \geq T\left(E(x,x),\mu(x)\right) = \mu(x)$.
3.

$$\begin{aligned}\phi_E(\mu \vee \mu')(x) &= \sup_{y\in X} T\left(E(x,y),(\mu \vee \mu')(y)\right)\\ &= \sup_{y\in X} T\left(E(x,y),\mu(y) \vee \mu'(y)\right)\\ &= \sup_{y\in X} T\left(E(x,y),\mu(y)\right) \vee \sup_{y\in X} T\left(E(x,y),\mu'(y)\right)\\ &= \phi_E(\mu)(x) \vee \phi_E(\mu')(x).\end{aligned}$$

4.

$$\begin{aligned}\phi_E(\phi_E(\mu))(x) &= \sup_{y\in X} T\left(E(x,y),\phi_E(\mu)(y)\right)\\ &= \sup_{y\in X}\sup_{z\in X} T\left(E(x,y),E(y,z),\mu(z)\right)\\ &\leq \sup_{z\in X} T\left(E(x,z),\mu(z)\right) = \phi_E(\mu)(x).\end{aligned}$$

   So $\phi_E(\phi_E(\mu)) \leq \phi_E(\mu)$. From 2 equality holds.
5. $\phi_E(\{x\})(y) = \sup_{z\in X} T\left(E(y,z),\{x\}(z)\right) = E(x,y)$ and the result follows from symmetry.
6.

$$\begin{aligned}\phi_E(T(\alpha,\mu(x)) &= \sup_{y\in X} T\left(E(x,y),T(\alpha,\mu)(y)\right)\\ &= \sup_{y\in X} T\left(E(x,y),\alpha,\mu(y)\right)\\ &= T(\alpha,\phi_E(\mu)(x)).\end{aligned}$$

There is a bijection between the operators satisfying Proposition 3.8 and T-indistinguishability operators.

Lemma 3.9. *Let ϕ be a map $\phi : [0,1]^X \to [0,1]^X$ satisfying the properties of Proposition 3.8. Then*

$$\phi(\mu)(x) = \sup_{y\in X} T(\phi(\{y\}),\mu(y))(x).$$

Proof. Clearly $\mu(x) = \sup_{y\in X} T(\{y\},\mu(y))(x)$.

By 3.8.3 and 3.8.6

$$\phi(\mu)(x) = \phi(\sup_{y\in X} T(\{y\},\mu(y))(x) = \sup_{y\in X} T(\phi(\{y\}),\mu(y))(x).$$

Proposition 3.10. *Let $\phi : [0,1]^X \to [0,1]^X$ be a map satisfying the properties of Proposition 3.8. The fuzzy relation E_ϕ on X defined for all $x, y \in X$ by*

$$E_\phi(x,y) = \phi(\{x\})(y)$$

is a T-indistinguishability operator on X.

Proof. For all $x, y, z \in X$

- Reflexivity: $E_\phi(x,x) = \phi(\{x\})(x) \geq \{x\}(x) = 1$.
- Symmetry follows from condition 3.8.5.
- T-transitivity:

$$\begin{aligned} E_\phi(x,y) &= \phi(\{x\})(y) = \phi(\phi(\{x\}))(y) \\ &= \sup_{u \in X} T(\phi(\{x\})(u), \phi(\{u\}))(y) \\ &= \sup_{u \in X} T(E_\phi(x,u), E_\phi(u,y)). \end{aligned}$$

Proposition 3.11. *There is a bijection between the set of T-indistinguishability operators and maps ϕ satisfying the conditions of Proposition 3.8.*

Proof. We have to prove that $\phi_{E_\phi} = \phi$ and $E_{\phi_E} = E$.

- Let ϕ be a map satisfying the conditions of Proposition 3.8. For all $x, y \in X$ and all fuzzy subsets μ of X we have

$$\phi_{E_\phi}(\mu)(x) = \sup_{y \in X} T(E_\phi(x,y), \mu(y)) = \sup_{y \in X} T(\phi(\{y\})(x), \mu(y)) = \phi(\mu)(x).$$

- Given a T-indistinguishability operator E, then for all $x, y \in X$ we have

$$E_{\phi_E}(x,y) = \phi(\{x\})(y) = \sup_{z \in X} T(E(y,z), \{x\}(z)) = E(x,y).$$

Proposition 3.12. $\mu \in H_E$ *if and only if* $\phi_E(\mu) = \mu$.

Proof. $\mu \in H_E$ if and only if $T(E(x,y), \mu(y)) \leq \mu(x)$ for all $x, y \in X$. Therefore $\phi_E(\mu) \leq \mu$. Since $\phi_E(\mu) \geq \mu$ holds for all fuzzy subsets, equality follows.

Hence, H_E is characterized as the set of fixed points of ϕ_E.

Proposition 3.13. $\text{Im}(\phi_E) = H_E$.

Proof.

- If $\mu \in H_E$, then $\phi_E(\mu) = \mu$ and hence $\mu \in \text{Im}(\phi_E)$.
- If $\mu = \phi_E(\nu)$, then $\phi_E(\mu) = \phi_E(\phi_E(\nu)) = \phi_E(\nu) = \mu$. Therefore $\mu \in H_E$.

Proposition 3.14. *For any* $\mu \in [0,1]^X$, $\phi_E(\mu) = \inf_{\mu' \in H_E}\{\mu \leq \mu'\}$.

Proof

$$\begin{aligned}\inf_{\mu' \in H_E}\{\mu \leq \mu'\} &= \inf_{\phi_E(\mu') \in H_E}\{\mu \leq \phi_E(\mu')\} \\ &= \inf_{\phi_E(\mu') \in H_E}\{\phi_E(\mu) \leq \phi_E(\mu')\} = \phi_E(\mu).\end{aligned}$$

So, $\phi_E(\mu)$ is the most specific extensional set that contains μ (i.e. $\mu \leq \phi_E(\mu)$) and in this sense it is the optimal upper bound of μ in H_E.

3.1.2 The Map ψ_E

Now, let us study the map ψ_E that maps each fuzzy subset to the greatest extensional fuzzy subset $\psi_E(\mu)$ contained in μ (i.e. $\psi_E(\mu) \leq \mu$).

Definition 3.15. *Let E be a T-indistinguishability operator on a set X. The map* $\psi_E : [0,1]^X \to [0,1]^X$ *is defined by*

$$\psi_E(\mu)(x) = \inf_{y \in X} \overrightarrow{T}(E(x,y)|\mu(y)) \ \forall x \in X.$$

Lemma 3.16. $\psi_E(\overrightarrow{T}(\{x\}|\alpha)(y) = \overrightarrow{T}(E(x,y)|\alpha)$.

Proof

$$\begin{aligned}\psi_E(\overrightarrow{T}(\{x\}|\alpha)(y) &= \inf_{z \in X} \overrightarrow{T}(E(y,z)|\overrightarrow{T}(\{x\}|\alpha)(z)) \\ &= \inf_{z \in X} \overrightarrow{T}(E(y,z)|\overrightarrow{T}(\{x\}(z)|\alpha)) \\ &= \min(\overrightarrow{T}(E(y,x)|\overrightarrow{T}(1|\alpha)), \inf_{z \in X, z \neq x} \overrightarrow{T}(E(y,z)|\overrightarrow{T}(0|\alpha))) \\ &= \min(\overrightarrow{T}(E(x,y)|\alpha), \inf_{z \in X, z \neq x} \overrightarrow{T}(E(y,z)|1)) \\ &= \min(\overrightarrow{T}(E(x,y)|\alpha), 1) = \overrightarrow{T}(E(x,y)|\alpha).\end{aligned}$$

Proposition 3.17. *For all* $\mu, \mu' \in [0,1]^X$, *we have:*

1. $\mu \leq \mu' \Rightarrow \psi_E(\mu) \leq \psi_E(\mu')$.
2. $\psi_E(\mu) \leq \mu$.
3. $\psi_E(\bigwedge_{i \in I} \mu_i) = \bigwedge_{i \in I} \psi_E(\mu_i)$.
4. $\psi_E(\psi_E(\mu)) = \psi_E(\mu)$.
5. $\psi_E(\overrightarrow{T}(\{x\}|\alpha))(y) = \psi_E(\overrightarrow{T}(\{y\}|\alpha))(x)$.
6. $\psi_E(\overrightarrow{T}(\alpha|\mu)) = \overrightarrow{T}(\alpha|\psi_E(\mu))$.

Proof

1. $\overrightarrow{T}$ is non-decreasing in the second variable.
2.

$$\begin{aligned}\psi_E(\mu)(x) &= \inf_{y\in X} \overrightarrow{T}(E(x,y)|\mu(y)) \\ &\leq \overrightarrow{T}(E(x,x)|\mu(x)) = \mu(x).\end{aligned}$$

3.

$$\begin{aligned}\psi_E(\bigwedge_{i\in I}\mu_i)(x) &= \inf_{y\in X} \overrightarrow{T}(E(x,y)|\bigwedge_{i\in I}\mu_i(y)) \\ &= \inf_{y\in X, i\in I} \overrightarrow{T}(E(x,y)|\mu_i(y)) = \bigwedge_{i\in I}\psi_E(\mu_i)(x).\end{aligned}$$

4.

$$\begin{aligned}\psi_E(\psi_E(\mu))(x) &= \inf_{y\in X} \overrightarrow{T}(E(x,y)|\psi_E(\mu)(y)) \\ &= \inf_{y\in X} \overrightarrow{T}(E(x,y)|\inf_{z\in X} \overrightarrow{T}(E(y,z)|\mu(z))) \\ &= \inf_{x,y\in X} \overrightarrow{T}(E(x,y)|\overrightarrow{T}(E(y,z)|\mu(z))) \\ &= \inf_{x,y\in X} \overrightarrow{T}(T(E(x,y),E(y,z))|\mu(z)) \\ &\geq \inf_{z\in X} \overrightarrow{T}(E(x,z)|\mu(z)) = \psi_E(\mu)(x).\end{aligned}$$

   So $\psi_E(\psi_E(\mu)) \geq \psi_E(\mu)$. From 2 we get the other inequality.
5. It follows from Lemma 3.16 and the symmetry of E.
6.

$$\begin{aligned}\psi_E(\overrightarrow{T}(\alpha|\mu))(x) &= \inf_{y\in X} \overrightarrow{T}(E(x,y)|\overrightarrow{T}(\alpha|\mu)(y)) \\ &= \inf_{y\in X} \overrightarrow{T}(E(x,y)|\overrightarrow{T}(\alpha|\mu(y))) \\ &= \inf_{y\in X} \overrightarrow{T}(\alpha|\overrightarrow{T}(E(x,y)|\mu(y))) \\ &= \overrightarrow{T}(\alpha|\inf_{y\in X} \overrightarrow{T}(E(x,y)|\mu(y))) \\ &= \overrightarrow{T}(\alpha|\psi_E(\mu)(x)) = \overrightarrow{T}(\alpha|\psi_E(\mu))(x).\end{aligned}$$

A similar result to Proposition 3.11 can be established between T-indistinguishability operators and maps satisfying the conditions of Proposition 3.17.

Proposition 3.18. *There is a bijection between the T-indistinguishability operators on X and the maps on X satisfying the conditions of Proposition 3.17.*

- *A T-indistinguishability operator E defines ψ_E which satisfies the conditions.*
- *From a map ψ satisfying the conditions we get the T-indistinguishability operator E_ψ defined for all $x, y \in X$ by*

$$E(x,y) = \inf_{\alpha\in[0,1]} \overrightarrow{T}(\psi(\overrightarrow{T}(\{x\}|\alpha)(y)|\alpha)).$$

Proposition 3.19. *$\mu \in H_E$ if and only if $\psi_E(\mu) = \mu$.*

Proof. $\mu \in H_E$ if and only if for all $x, y \in X$ $T(E(x,y), \mu(x)) \leq \mu(y)$, which is equivalent to $\overrightarrow{T}(E(x,y)|\mu(y)) \geq \mu(x)$ for all $x, y \in X$. This again is equivalent to $\inf_{y\in X} \overrightarrow{T}(E(x,y)|\mu(y)) \geq \mu(x)$. In other words, $\psi_E(\mu)(x) \geq \mu(x)$.

So, H_E is also characterized as the set of fixed points of ψ_E.

Proposition 3.20. $\mathrm{Im}(\psi_E) = H_E$.

Proof

- If $\mu \in H_E$, then $\psi_E(\mu) = \mu$. Therefore $\mu \in \mathrm{Im}(\psi_E)$.
- If $\mu \in \mathrm{Im}\psi_E$, then there exists ν with $\psi_E(\nu) = \mu$. So $\mu = \psi_E(\nu) = \psi_E(\psi_E(\nu)) = \psi_E(\mu)$ and $\mu \in H_E$.

Proposition 3.21. *For any $\mu \in [0,1]^X$, $\psi_E(\mu) = \sup_{\mu'\in H_E}\{\mu' \leq \mu\}$.*

Proof

$$\begin{aligned}\sup_{\mu'\in H_E}\{\mu' \leq \mu\} &= \sup_{\psi_E(\mu')\in H_E}\{\psi_E(\mu') \leq \mu\} \\ &= \sup_{\psi_E(\mu')\in H_E}\{\psi_E(\mu') \leq \psi_E(\mu)\} \\ &= \psi_E(\mu).\end{aligned}$$

3.1.3 Properties of ϕ_E and ψ_E. Fuzzy Rough Sets and Modal Logic

The maps ϕ_E and ψ_E provide upper and lower approximations to any fuzzy subset μ of X. This is the key idea of rough sets. Lets recall that if there is a crisp equivalence relation $\sim$ defined on a set X, a crisp subset $A \subseteq X$ is approximated by

$$\overline{A} = \{x \in X \text{ such that } \overline{x} \cap A \neq \emptyset\}$$

and

$$\underline{A} = \{x \in X \text{ such that } \overline{x} \subseteq A\}$$

where $\overline{x}$ denotes the equivalence class of x. $(\overline{A}, \underline{A})$ is called a rough set [110] .

In this way, every crisp set is approximated by the closest sets that can be observed when the equivalence relation $\sim$ is taken into account. Let us note that the definitions of the approximations $\overline{A}$ and $\underline{A}$ can also be written

$$x \in \overline{A} \text{ if and only if } \exists y \in X \text{ such that } y \in A \wedge x \sim y$$

and

$$x \in \underline{A} \text{ if and only if } \forall y \in X \; x \sim y \rightarrow y \in A.$$

which are $\phi_\sim(A)$ and $\psi_\sim(A)$.

Therefore ϕ_E and ψ_E generalize rough sets to fuzzy rough fuzzy sets.

ϕ_E and ψ_E have also been applied to model the fuzzy possibility and the fuzzy necessity operators in fuzzy modal logic, where the accessibility relation between possible worlds is given by a T-indistinguishability operator E or, more general, by a fuzzy relation [21].

In general a fuzzification of a Kripke frame is a pair $F = (W, R)$ where W is the set of possible worlds and R a fuzzy relation on W called the accessibility relation.

A Kripke model is a 3-tuple $M = (W, R, V)$ where (W, R) is a Kripke frame and V is a map, called valuation, assigning to each variable in Var and each world in W an element of $[0, 1]$ (i.e., $V : Var \times W \rightarrow A$).

The valuation V can be extended to any formula φ and given a formula φ the map $V_\varphi : W \rightarrow [0, 1]$ can be defined as $V_\varphi(w) = V(\varphi, w)$.

Then valuation of the necessity $\Box\varphi$ of φ in world w is then defined by

$$V(\Box\varphi, w) = \psi_R(V_\varphi)(w) = \inf_{w' \in W} \overrightarrow{T}\,(R(w, w')|V_\varphi(w'))$$

and the possibility $\Diamond\varphi$ of φ in world w is

$$V(\Diamond\varphi, w) = \phi_R(V_\varphi)(w) = \sup_{w' \in W} T(R(w, w'), V_\varphi(w')).$$

In a classical setting, the possibility is often defined from the necessity by $\Diamond = \neg\Box\neg$ where $\neg$ is the negation. In the fuzzy case we do not have this equality in general. In order to study the cases when $\phi_E = \varphi \circ \psi_E \circ \varphi$ where φ is a strong negation is satisfied we need first to recall the characterization theorem for strong negations.

Definition 3.22. [134] *A strong negation φ is defined as a strictly decreasing, continuous function $\varphi : [0, 1] \rightarrow [0, 1]$ with boundary conditions $\varphi(0) = 1$ and $\varphi(1) = 0$ such that φ is involutive (i.e., $\varphi(\varphi(x)) = x$ holds for any $x \in [0, 1]$).*

The standard strong negation is $\varphi(x) = 1 - x$.

Theorem 3.23. [134] *A map $\varphi : [0, 1] \rightarrow [0, 1]$ is a strong negation if and only if there exists a continuous and decreasing map $t : [0, 1] \rightarrow \mathbb{R}^+$ with $t(1) = 0$ such that*

$$\varphi(x) = t^{-1}(t(0) - t(x)) \ \forall x \in [0, 1].$$

A map t as defined in the preceding theorem is called a generator of the strong negation φ.

This theorem can be interpreted in the following way.

Corollary 3.24. *φ is a strong negation if and only if there exists a continuous non-strict Archimedean t-norm T with additive generator t such that*

$$\varphi(x) = \overrightarrow{T}(x|0).$$

Proposition 3.25. *Let X be a set and φ a strong negation with generator t. If T is a non-strict Archimedean t-norm with additive generator t and E is a T-indistinguishability operator on X, then $\varphi \circ \phi_E \circ \varphi = \psi_E$ (and therefore $\varphi \circ \psi_E \circ \varphi = \phi_E$).*

Proof

$$\begin{aligned}
\psi_E(\mu)(x) &= \inf_{y \in X} \overrightarrow{T}(E(x, y)|\mu(y)) \\
&= \inf_{y \in X} (t^{[-1]}(t(\mu(y)) - t(E(x, y)) \\
&= \inf_{y \in X} t^{[-1]}(t(0) - (t(E(x, y)) + t(0) - t(\mu(y)))) \\
&= \inf_{y \in X} t^{[-1]}(t(0) - t(t^{[-1]}(t(E(x, y)) + t(t^{[-1]}(t(0) - t(\mu(y)))))) \\
&= \inf_{y \in X} \varphi(T(E(x, y), \varphi(\mu(y)))) \\
&= \varphi(\sup_{y \in X} T(E(x, y), \varphi(\mu(y)))) \\
&= \varphi(\phi_E(\varphi(\mu)(x))).
\end{aligned}$$

The following reciprocal result holds.

Proposition 3.26. *Let X be a set, T a continuous Archimedean t-norm, φ a strong negation and E a non crisp T-indistinguishability operator on X. If $\varphi \circ \phi_E \circ \varphi = \psi_E$, then T is a non-strict Archimedean t-norm and $\varphi_{|D} = \varphi_{T|D}$ where $D = \{E(x, y)$ with $x, y \in X\}$ and φ_T is the strong negation associated to T ($\varphi_T(x) = \overrightarrow{T}(x|0)$).*

Proof. Let us first suppose that T is a strict Archimedean t-norm with generator t. Let $x, y \in X$ be such that $E(x, y) \notin \{0, 1\}$ and consider the column μ_x of E which is an extensional fuzzy subset ($\mu_x \in H_E$).

We want to prove that $\psi_E(\mu_x) \neq \varphi \circ \phi_E \circ \varphi(\mu_x)$, which is equivalent to prove $\varphi \circ \psi_E(\mu_x) \neq \phi_E \circ \varphi(\mu_x)$.

Since $\phi_E \circ \varphi(\mu_x) \in H_E$, to prove the last inequality it suffices to see that $\varphi \circ \psi_E(\mu_x) \notin H_E$. But since $\psi_E(\mu_x) = \mu_x$, this is equivalent to see that $\varphi(\mu_x) \notin H_E$.

$$\begin{aligned}E_{\varphi(\mu_x)}(x,y) &= E_T(\varphi(E(x,x)),\varphi(E(x,y)))\\ &= \overrightarrow{T}(\varphi(E(x,y))|0) = 0 < E(x,y)\end{aligned}$$

and therefore $\varphi(\mu_x) \notin H_E$.

On the other hand, since μ_x is extensional with respect to E,

$$\begin{aligned}\varphi(\mu_x(z)) &= \varphi(\psi_E(\mu_x(z)))\\ &= \phi_E(\varphi(\mu_x)(z))\\ &= \sup_{y\in X} T(E(z,y),\varphi(\mu_x(y)))\\ &= \sup_{y\in X} t^{[-1]}(t(E(z,y))+t(\varphi(E(x,y)))).\end{aligned}$$

In particular,

$$\varphi(\mu_x(x)) = 0 = \sup_{y\in X} t^{[-1]}(t(E(x,y))+t(\varphi(E(x,y)))).$$

Therefore, for all $y \in X$, $t(E(x,y)) + t(\varphi(E(x,y))) = t(0)$ and

$$\varphi(E(x,y)) = t^{[-1]}(t(0)-t(E(x,y))) = \varphi_T(E(x,y)).$$

Let us end this subsection with some properties combining ϕ_E and ψ_E to show their dual behaviour.

Proposition 3.27. *Let E be a T-indistinguishability operator on a set X and μ a fuzzy subset of X.*

1. $\psi_E(\mu) = \inf_{\alpha\in[0,1]} \overrightarrow{T}(\phi_E(\overrightarrow{T}(\mu|\alpha))|\alpha)$.
2. $\phi_E(\mu) = \inf_{\alpha\in[0,1]} \overrightarrow{T}(\psi_E(\overrightarrow{T}(\mu|\alpha))|\alpha)$.

Proof

1.

$$\begin{aligned}\inf_{\alpha\in[0,1]} \overrightarrow{T}(\phi_E(\overrightarrow{T}(\mu(x)|\alpha))|\alpha) &= \inf_{\alpha\in[0,1]} \overrightarrow{T}(\sup_{y\in X} T(E(y,x),\overrightarrow{T}(\mu(y)|\alpha))|\alpha)\\ &= \inf_{\alpha\in[0,1]}\inf_{y\in X} \overrightarrow{T}(T(E(y,x),\overrightarrow{T}(\mu(y)|\alpha))|\alpha)\\ &= \inf_{y\in X}\inf_{\alpha\in[0,1]} \overrightarrow{T}(E(y,x)|\overrightarrow{T}(\overrightarrow{T}(\mu(y)|\alpha))|\alpha)\\ &= \inf_{y\in X} \overrightarrow{T}(E(y,x)|\inf_{\alpha\in[0,1]} \overrightarrow{T}(\overrightarrow{T}(\mu(y)|\alpha))|\alpha)\\ &= \inf_{y\in X} \overrightarrow{T}(E(y,x)|\mu(y))\\ &= \psi_E(\mu)(x).\end{aligned}$$

2. can be proved in a similar way.

Proposition 3.28. *For every $\alpha \in [0,1]$ and μ, ν fuzzy subsets of X,*

1. $\phi_E(\psi_E(\mu)) = \psi_E(\mu)$.
2. $\psi_E(\phi_E(\mu)) = \phi_E(\mu)$.
3. $\psi_E(\overrightarrow{T}(\mu|\alpha)) = \overrightarrow{T}(\phi_E(\mu)|\alpha)$.
4. $\inf_{x\in X} \overrightarrow{T}(\mu(x)|\psi_E(\nu)(x)) = \inf_{x\in X} \overrightarrow{T}(\phi_E(\mu)(x)|\nu(x))$.
5. $\inf_{x\in X} \overrightarrow{T}(\psi_E(\mu)(x)|\psi_E(\nu)(x)) = \inf_{x\in X} \overrightarrow{T}(\psi_E(\mu)(x)|\nu(x))$.
6. $\inf_{x\in X} \overrightarrow{T}(\phi_E(\mu)(x)|\phi_E(\nu)(x)) = \inf_{x\in X} \overrightarrow{T}(\mu(x)|\phi_E(\nu)(x))$.

Proof

1. $\psi_E(\mu)$ is extensional.
2. $\phi_E(\mu)$ is extensional.
3.

$$\begin{aligned}\psi_E(\overrightarrow{T}(\mu|\alpha)(x)) &= \inf_{y\in X} \overrightarrow{T}(E(y,x)|\overrightarrow{T}(\mu(y)|\alpha)) \\ &= \inf_{y\in X} \overrightarrow{T}(T(E(y,x),\mu(y))|\alpha) \\ &= \overrightarrow{T}(\sup_{y\in X} T(E(y,x),\mu(y))|\alpha) \\ &= \overrightarrow{T}(\phi_E(\mu)(x)|\alpha).\end{aligned}$$

4.

$$\begin{aligned}\inf_{x\in X} \overrightarrow{T}(\mu(x)|\psi_E(\nu)(x)) &= \inf_{x\in X} \overrightarrow{T}(\mu(x)|\inf_{y\in X} \overrightarrow{T}(E(y,x)|\nu(x))) \\ &= \inf_{x\in X}\inf_{y\in X} \overrightarrow{T}(T(E(y,x)|\mu(x))|\nu(y)) \\ &= \inf_{y\in X} \overrightarrow{T}(\sup_{x\in X} T(E(y,x),\mu(x))|\nu(y)) \\ &= \inf_{x\in X} \overrightarrow{T}(\phi_E(\mu)(x)|\nu(x)).\end{aligned}$$

5.

$$\begin{aligned}\inf_{x\in X} \overrightarrow{T}(\psi_E(\mu)(x)|\psi_E(\nu)(x)) &= \inf_{x\in X} \overrightarrow{T}(\phi_E(\psi_E(\mu))(x)|\nu(x)) \\ &= \inf_{x\in X} \overrightarrow{T}(\psi_E(\mu)(x)|\nu(x)).\end{aligned}$$

6.

$$\begin{aligned}\inf_{x\in X} \overrightarrow{T}(\phi_E(\mu)(x)|\phi_E(\nu)(x)) &= \inf_{x\in X} \overrightarrow{T}(\mu(x)|\psi_E(\phi_E(\mu))(x)) \\ &= \inf_{x\in X} \overrightarrow{T}(\mu(x)|\phi_E(\nu)(x)).\end{aligned}$$

3.1.4 Topological Structure of H_E

As we will see in this subsection, the set H_E of extensional sets of a T-indistinguishability operator on a set X is the set of closed fuzzy subsets of a fuzzy topology defined on X and the maps ϕ_E and ψ_E are fuzzy closure and interior operators of it. E by its side also generates a classical topology on X and the tight relation between both classical and crisp topologies will be established.

Let us recall the definitions of fuzzy topology, fuzzy closure operator and fuzzy interior operator ([91]).

Definition 3.29. *A subset τ of $[0,1]^X$ is a fuzzy topology on X if and only if it satisfies the following properties.*

1. *If $A \subseteq \tau$, then* $\sup\{\mu \text{ such that } \mu \in A\} \in \tau$.
2. *If $\mu, \mu' \in \tau$, then* $\min(\mu, \mu') \in \tau$.
3. *If μ_k is a constant fuzzy subset of X (i.e. $\mu_k(x) = k \in [0,1]$ for all $x \in X$), then $\mu_k \in \tau$.*

Definition 3.30. *A map $\phi : [0,1]^X \to [0,1]^X$ is a fuzzy closure operator if and only if for all fuzzy subsets μ, μ' of X satisfies*

1. $\mu \leq \phi(\mu)$.
2. $\phi(\mu \vee \mu') = \phi(\mu) \vee \phi(\mu')$.
3. $\phi(\phi((\mu)) = \phi(\mu)$.
4. $\phi(\mu_k) = \mu_k$ *for every constant fuzzy subset μ_k of X.*

Definition 3.31. *A map $\psi : [0,1]^X \to [0,1]^X$ is a fuzzy interior operator if and only if for all fuzzy subsets μ, μ' of X satisfies*

1. $\psi(\mu) \leq \mu$.
2. $\psi(\mu \wedge \mu') = \psi(\mu) \wedge \psi(\mu')$.
3. $\psi(\psi(\mu)) = \psi(\mu)$.
4. $\psi(\mu_k) = \mu_k$ *for every constant fuzzy subset μ_k of X.*

The elements of a fuzzy topology τ are called open fuzzy subsets. Given a strong negation φ a fuzzy subset μ is called closed if and only if $\varphi \circ \mu$ is open.

It is easy to prove that, given a fuzzy interior operator ψ on a set X, the fuzzy subsets μ of X satisfying $\psi(\mu) = \mu$ (the fixed points of ψ) are the open subsets of a fuzzy topology τ_ψ. Also the fixed points of a fuzzy closure operators are the closed subsets of a fuzzy topology of X.

Some of the results of the previous two subsections can be translated to the following proposition.

Proposition 3.32. *Let E be a T-indistinguishability operator on a set X.*

- *The map ϕ_E is a fuzzy closure operator.*
- *The map ψ_E is a fuzzy interior operator.*

- *The set H_E of extensional sets of E is a fuzzy topology on X and also the set of closed fuzzy subsets of a topology on X.*

On the other hand, a T-indistinguishability operator E on a set X defines a crisp topology on X in a natural way.

Definition 3.33. *Given a T-indistinguishability operator E separating points on a set X and a strong negation φ, the open ball $B(x,r)$ of centre $x \in X$ and radius $r \in (0,1]$ is defined by*

$$B(x,r) = \{y \in X \text{ such that } E(x,y) > \varphi(r)\}.$$

The set of open balls is a basis of a topology denoted by T_E.

In order to relate the topologies T_E and τ_E generated by a T indistinguishability operator E, we recall the following definitions ([91]).

Definition 3.34. *Let $T(X)$ and $\tau(X)$ denote the sets of topologies and fuzzy topologies on a given set X respectively. The mappings $i : \tau(X) \to T(X)$ and $\omega : T(X) \to \tau(X)$ are defined in the following way.*

$$i(\tau) = \left\{\mu^{-1}(\alpha, 1] \text{ such that } \alpha \in [0,1), \mu \in \tau\right\}.$$
$$\omega(T) = \left\{\mu \in [0,1]^X \text{ such that } \mu \text{ is lower semicontinuous}\right\}.$$

For these two maps we have

$$\omega \circ i = id_{\tau(X)}.$$

Proposition 3.35. *If T and τ are the topology and the fuzzy topology generated on a set X by a T-indistinguishability operator E separating points and φ is a strong negation, then $i(\tau) = T$ (and therefore $\omega(T) = \tau$).*

Proof

- $i(\tau) \subseteq T$: If $A \in i(\tau)$, then there exist $\alpha \in [0,1)$ and $\mu \in \tau$ such that $A = \mu^{-1}(\alpha, 1] = \{x \in X \text{ such that } \mu(x) > \alpha\}$. If $A = \emptyset$, then $A \in T$. Otherwise, for each $x \in A$, the ball of centre x and radius $r = \varphi\left(\overrightarrow{T}(\varphi(\alpha)|\varphi(\mu(x)))\right)$ is contained in A. Indeed, given $y \in B(x,r)$,

  1. if $\mu(y) \geq \mu(x)$, then $\mu(y) > \alpha$.
  2. if $\mu(y) < \mu(x)$, then

  $$\begin{aligned} E_{\varphi\circ\mu}(x,y) &= \overrightarrow{T}(\varphi(\mu(y))|\varphi(\mu(x))) \\ &\geq E(x,y) > \overrightarrow{T}(\varphi(\alpha)|\varphi(\mu(x))). \end{aligned}$$

  Since $\overrightarrow{T}$ is decreasing in its first component, $\varphi(\mu(y)) < \varphi(\alpha)$ and therefore $\mu(y) > \alpha$.

- $T \subseteq i(\tau)$: Let us prove that any given ball $B(x,r)$ of T is an open set of $i(\tau)$.

  If $\mu_{x,r}$ is defined by $\mu_{x,r} = \varphi(\overrightarrow{T}(E(x,y)|\varphi(r))$, then

$$B(x,r) = \{y \in X \text{ such that } E(x,y) > \varphi(r)\}$$
$$= \{y \in X \text{ such that } \overrightarrow{T}(E(x,y)|\varphi(r)) < 1\} = \mu_{x,r}^{-1}(0,1].$$

Therefore it is sufficient to prove that $\varphi(\mu_{x,r}) \in H_E$. Indeed, given $y, z \in X$ we can assume without loss of generality that $\mu(y) \geq \mu(z)$ and

$$E_{\varphi(\mu)}(x,y) = \overrightarrow{T}(\overrightarrow{T}(E(x,y)|\varphi(r))|\overrightarrow{T}(E(x,z)|\varphi(r)))$$
$$\geq \overrightarrow{T}(E(x,z)|E(x,y)) \geq E(y,z).$$

3.2 Fuzzy Points and the Map Λ_E

In the present section, we are going to associate a new map Λ_E to a given T-indistinguishability operator E, which is also closely related to the structure of E. The main result concerning Λ_E is that it has the columns of E as fixed points.

The set $\text{Im}(\Lambda_E)$ will be characterized as the set of fixed points of Λ_E^2. In this way, $\text{Im}(\Lambda_E)$ appears as a well differentiated subset of H_E. $\text{Fix}(\Lambda_E)$ will be characterized as the set of maximal fuzzy points of E.

Definition 3.36. *Let E be a T-indistinguishability operator on a set X. $\mu \in H_E$ is a fuzzy point of X with respect to E if and only if*

$$T(\mu(x_1), \mu(x_2)) \leq E(x_1, x_2) \ \forall x_1, x_2 \in X.$$

P_X will denote the set of fuzzy points of X with respect to E.

Definition 3.37. *Let $E, \overline{E}$ be two T-indistinguishability operators on X and $\overline{X}$ respectively. $(\overline{X}, \overline{E})$ is an extension of (X, E) if and only if*

1. *$X \subseteq \overline{X}$*
2. *$\overline{E}(x,y) = E(x,y) \ \forall x, y \in X$.*

More general,

Definition 3.38. *Let E, F be T-indistinguishability operators on X and Y respectively. A map $\tau : X \to Y$ is an isometric embedding of (X, E) into (Y, F) if and only if*

$$E(x,y) = F(\tau(x), \tau(y)) \ \forall x, y \in X.$$

The next Theorem provides us with a criterion to decide whether an extensional set is a fuzzy point.

Theorem 3.39. *Let E be a T-indistinguishability operator on X. Given $\mu \in H_E$, these are equivalent statements:*

a) μ is a fuzzy point.
b) There exists an extension $(\overline{X}, \overline{E})$ of (X, E) such that $\mu = \mu_y|_X$ for a particular $y \in \overline{X}$ (i.e. $\mu(x) = \overline{E}(y, x)\ \forall x \in X$).

Proof. b) $\Rightarrow$ a)

$$\begin{aligned} T(\mu(x_1), \mu(x_2)) &= T(\overline{E}(y, x_1), \overline{E}(y, x_2)) \\ &\leq \overline{E}(x_1, x_2) = E(x_1, x_2) \end{aligned}$$

for all $x_1, x_2 \in X$.

a) $\Rightarrow$ b)

We define a T-indistinguishability operator $\overline{E}$ on the set $\overline{X} = X \cup \{\mu\}$ as follows:

$$\begin{aligned} \overline{E}(x_1, x_2) &= E(x_1, x_2)\ \forall x_1, x_2 \in X \\ \overline{E}(x, \mu) &= \overline{E}(\mu, x) = \mu(x)\ \forall x \in X \\ \overline{E}(\mu, \mu) &= 1. \end{aligned}$$

$\overline{E}$ is reflexive and symmetric and it is an extension of E.

It remains to prove the T-transitivity of $\overline{E}$, i.e. $T(\overline{E}(x, y), \overline{E}(y, z)) \leq \overline{E}(x, z)$. There are only four possible (non exclusive) cases:

- $x = y$, $y = z$ or $x = z$ (trivial)
- $x, y, z \in X$ (trivial)
- $y = \mu$ and $x, z \in X$. In this case

$$T(\overline{E}(x, \mu), \overline{E}(\mu, z)) = T(\mu(x), \mu(z)) \leq E(x, z).$$

- $x = \mu$ and $y, z \in X$. In this case

$$T(\overline{E}(\mu, y), \overline{E}(y, z)) = T(\mu(y), E(y, z)) \leq \mu(z) = \overline{E}(\mu, z),$$

because $\mu \in H_E$.

This theorem characterizes both the columns of E and the columns of its extensions as exactly the fuzzy points of E. We note

$$\begin{aligned} C_E = \{\mu \in H_E\ | \exists (\overline{X}, \overline{E}) \text{ extension of } (X, E) \text{ and } y \in \overline{X} \text{ such that} \\ \mu(x) = \overline{E}(y, x),\ \forall x \in X\}. \end{aligned}$$

Of course, $P_X = C_E$ and we will say that a fuzzy point is in C_E when we want to stress the idea that it can be a column of an extension of (X, E).

If μ is normal, then $(\overline{X}, \overline{E}) = (X, E)$, and $\mu = \mu_x$ for some $x \in X$. This particular case is a well known result (see, for example, [23]).

In order to have a characterization of the columns of E, let us introduce the map Λ_E [20].

Definition 3.40. *Let E be a T-indistinguishability operator on a set X. The map $\Lambda_E : [0,1]^X \to [0,1]^X$ is defined by*

$$\Lambda_E(\mu)(x) = \inf_{y \in X} \overrightarrow{T}(\mu(y)|E(y,x)) \;\; \forall x \in X.$$

It is easy to check that if E is a crisp equivalence relation $\sim$, then Λ_E acts simply by intersecting equivalence classes: $\Lambda_\sim(A) = \bigcap_{x \in A} \overline{x}$ where $\overline{x}$ is the equivalence class of x with respect to $\sim$. So that in a crisp framework only three different situations may occur. Namely:

- $A \neq \emptyset$ and there exists $x \in X$ such that $A \subseteq \overline{x}$. In this case, $\Lambda_\sim(A) = \overline{x}$. ($\Lambda_\sim(A)$ is the intersection of exactly one equivalence class $\overline{x}$).
- $\Lambda_\sim(A) = \emptyset$ in any other situation with $A \neq \emptyset$ ($\Lambda_\sim(A)$ is then the intersection of two or more equivalence classes).
- $\Lambda_\sim(\emptyset) = X$ (Note that $\emptyset \subseteq \overline{x}$ for all $x \in X$).

In other words, if a crisp subset A of X is contained in exactly one equivalence class $\overline{x}$ of $\sim$, then $\Lambda_\sim(A) = \overline{x}$. If A intersects more than an equivalence class of E, then $\Lambda_\sim(A) = \emptyset$ and $\Lambda_\sim(\emptyset) = X$.

This summarizes the situation in the crisp case. However, not such a trivial discussion can give understanding enough in the fuzzy case, mainly due to two reasons.

- First, there exist columns μ_y having their centers or prototypical elements y outside X (as it states Theorem 3.39).
- Second, the map Λ_E^2 (which in the crisp case is a trivial one, fixing the columns and sending X to $\emptyset$ and $\emptyset$ to X) plays here an important role as will be seen later on in this section.

Some general properties concerning Λ_E are:

Proposition 3.41. *Given $\mu_1, \mu_2 \in [0,1]^X$, we have:*

a) $\Lambda_E(\mu_1) \geq \Lambda_E(\mu_2)$ *if* $\mu_1 \leq \mu_2$.
b) $\Lambda_E(\mu_1 \vee \mu_2) = \Lambda_E(\mu_1) \wedge \Lambda_E(\mu_2)$.
c) $\Lambda_E(\mu_1 \wedge \mu_2) \geq \Lambda_E(\mu_1) \vee \Lambda_E(\mu_2)$.

Proof. Trivial.

Proposition 3.42. *For $\mu \in [0,1]^X$ and $\alpha \in [0,1]$,*

a) $\Lambda_E(T(\alpha, \mu)) = \overrightarrow{T}(\alpha|\Lambda_E(\mu))$.
b) $\Lambda_E(\overrightarrow{T}(\alpha|\mu)) \geq T(\alpha, \Lambda_E(\mu))$.

Proof. a)

$$\begin{aligned}\Lambda_E(T(\alpha,\mu))(x) &= \inf_{y\in X} \overrightarrow{T}(T(\alpha,\mu(y))|E(y,x))\\ &= \overrightarrow{T}(\alpha|\inf_{y\in X}\overrightarrow{T}(\mu(y)|E(y,x)))\\ &= \overrightarrow{T}(\alpha|\Lambda_E(\mu)(x))\end{aligned}$$

for all $x \in X$.

b)

$$\begin{aligned}\Lambda_E(\overrightarrow{T}(\alpha|\mu))(x) &= \inf_{y\in X} \overrightarrow{T}(\overrightarrow{T}(\alpha|\mu(y))|E(y,x))\\ &\geq T(\alpha,\inf_{y\in X}\overrightarrow{T}(\mu(y)|E(y,x)))\\ &= \overrightarrow{T}(\alpha|\Lambda_E(\mu)(x))\end{aligned}$$

for all $x \in X$.

The following two propositions establish the relation between $\text{Fix}(\Lambda_E)$ (the set of fixed points of Λ_E) and the columns of E.

Proposition 3.43. $\text{Fix}(\Lambda_E) \subseteq C_E = P_X$.

Proof. Let $\mu \in [0,1]^X$ be a fixed point of Λ_E, i.e. $\Lambda_E(\mu) = \mu$.

Being $\Lambda_E(\mu)(x) = \inf_{y\in X} \overrightarrow{T}(\mu(y)|E(x,y))$, then $\Lambda_E(\mu) = \mu$ implies that $\overrightarrow{T}(\mu(y)|E(x,y)) \geq \mu(x)$ for all $y \in X$ or, equivalently, $T(\mu(x),\mu(y)) \leq E(x,y)$ for all $y \in X$.

On the other hand, $\Lambda_E(h) \in H_E$ (see Proposition 3.47).

The set $\text{Fix}(\Lambda_E)$ will be characterized as the set of maximal elements of C_E in Theorem 3.58.

Proposition 3.44. *Let μ be a normal fuzzy subset of X (i.e. $\exists x_0 \in X$ such that $\mu(x_0) = 1$) $\Lambda_E(\mu) = \mu$ if and only if μ is a column μ_x of E.*

Proof. If $\Lambda_E(\mu) = \mu$, then $\mu \in C_E$ (Proposition 3.43) and being μ a normal fuzzy subset, we have $\mu = \mu_x$, for some $x \in X$.

Conversely, if $\mu = \mu_x$ for some $x \in X$, then using Lemma 2.39

$$\begin{aligned}\Lambda_E(\mu_x)(y) &= \inf_{z\in X} \overrightarrow{T}(\mu_x(z)|E(z,y))\\ &= \inf_{z\in X} \overrightarrow{T}(E(z,x)|E(z,y))\\ &= E(x,y) = \mu_x(y)\end{aligned}$$

for all $y \in X$.

Proposition 3.44 characterizes only the columns of elements $x \in X$ and it cannot be extended to the whole set C_E as it is shown in next example.

Example 3.45. $X = \{x_1, x_2\}$, $E(x_1, x_2) = 0$, T an arbitrary t-norm. We define the following extension of (X, E) : $\overline{X} \neq X \cup \{y\}$, $\overline{E}(x_1, y) = \overline{E}(x_2, y) = 0$.

The column μ_Y of y is (restricted to X), the constant fuzzy set $\mu_y(x_1) = \mu_y(x_2) = 0$. So that $\Lambda_E(\mu_y) = X$ i.e. $\Lambda_E(\mu_y)(x_1) = \Lambda_E(\mu_y)(x_2) = 1$.

However, there are also fixed points of Λ_E that are not columns μ_x, $x \in X$.

Example 3.46. For a given $n \in \mathbb{N}$, $n \geq 2$, let us consider $X = \{0, \frac{1}{n}, \frac{2}{n}, ..., \frac{n-1}{n}, 1\} \subseteq [0,1]$, T the Łukasiewicz t-norm and E defined by $E(x, y) = 1 - |x - y|$ for all $x, y \in X$.

Let μ be the non-normal fuzzy subset defined by $\mu(x) = 1 - \left|\frac{3}{2n} - x\right|$, $x \in X$. Obviously $\mu \neq \mu_x$ for all $x \in X$, and it is easy to check that $\Lambda_E(\mu) = \mu$.

Let us now turn our attention to $\mathrm{Im}(\Lambda_E)$. The map Λ_E^2 will play an essential role and the main result concerning it will identify its fixed points with the image of Λ_E.

Let us start by noting that $\Lambda_E(\mu)$ is always an extensional fuzzy subset, for any $\mu \in [0,1]^X$.

Proposition 3.47. $\mathrm{Im}(\Lambda_E) \subseteq H_E$.

Proof. For any $\mu \in [0,1]^X$, we have to prove that $\Lambda_E(\mu) \in H_E$.

$$\begin{aligned}
\overrightarrow{T}(\Lambda_E(\mu)(x_1)|\Lambda_E(\mu)(x_2)) &= \overrightarrow{T}(\inf_{y \in X} \overrightarrow{T}(\mu(y)|E(y, x_1))| \inf_{z \in X} \overrightarrow{T}(\mu(z)|E(z, x_2))) \\
&= \inf_{z \in X} \overrightarrow{T}(\inf_{y \in X} \overrightarrow{T}(\mu(y)|E(y, x_1))|\overrightarrow{T}(\mu(z)|E(z, x_2))) \\
&\geq \inf_{z \in X} \overrightarrow{T}(\overrightarrow{T}(\mu(z)|E(z, x_1))|\overrightarrow{T}(\mu(z)|E(z, x_2))) \\
&\geq \inf_{z \in X} \overrightarrow{T}(E(x_1, y)\ |E(x_2, y)) = E(x_1, x_2).
\end{aligned}$$

(Applying Lemmas 2.38, 2.39 and the T-transitivity of E).

In a similar way, we obtain $\overrightarrow{T}(\Lambda_E(\mu)(x_2)|\Lambda_E(\mu)(x_1)) \geq E(x_1, x_2)$, and therefore $E_T(\Lambda_E(\mu)(x_1), \Lambda_E(\mu)(x_2)) \geq E(x_1, x_2)$, for all $x_1, x_2 \in X$, so that $\Lambda_E(\mu) \in H_E$.

At this point, it is not clear whether the set $\mathrm{Im}(\Lambda_E)$ coincides with H_E or, on the contrary, it is strictly contained in H_E.

To answer this question, we turn out our attention to the operator Λ_E^2.

Proposition 3.48. *Given* $\mu_1, \mu_2 \in [0,1]^X$,

a) If $\mu_1 \leq \mu_2$ *then* $\Lambda_E^2(\mu_1) \leq \Lambda_E^2(\mu_2)$
b) $\Lambda_E^2(\mu_1 \vee \mu_2) \geq \Lambda_E^2(\mu_1) \vee \Lambda_E^2(\mu_2)$
c) $\Lambda_E^2(\mu_1 \wedge \mu_2) \leq \Lambda_E^2(\mu_1) \wedge \Lambda_E^2(\mu_2)$.

Proof. Trivial.

Proposition 3.49. $\Lambda_E^2 \geq \phi_E$

Proof. Given $\mu \in [0,1]^X$, we have:

$$\begin{aligned}
\Lambda_E^2(\mu)(x) &= \Lambda_E(\Lambda_E(\mu))(x) \\
&= \inf_{y\in X} \overrightarrow{T}(\Lambda_E(\mu)(y) \mid E(y,x)) \\
&= \inf_{y\in X} \overrightarrow{T}(\inf_{z\in X} \overrightarrow{T}(\mu(z)|E(z,y))|E(y,x)) \\
&\geq \inf_{y\in X} \sup_{z\in X} \overrightarrow{T}(\overrightarrow{T}(\mu(z)|E(z,y))|E(y,x)) \\
&\geq \sup_{z\in X} \inf_{y\in X} \overrightarrow{T}(\overrightarrow{T}(\mu(z)|E(z,y))|E(y,x)) \\
&\geq \sup_{z\in X} T(\mu(z), \inf_{y\in X} \overrightarrow{T}(E(y,z)|E(y,x))) \\
&= \sup_{z\in X} T(\mu(z), E(z,x)) = \phi_E(\mu)(x),
\end{aligned}$$

for all $x \in X$.

Corollary 3.50. *$\Lambda_E^2(\mu) \geq \mu$, for all $\mu \in [0,1]^X$.*

Proof. $\Lambda_E^2(\mu) \geq \phi_E(\mu) \geq \mu$.

Lemma 3.51. *Given a column μ_x, $x \in X$ and $\alpha \in [0,1]$ we have:*

a) $\Lambda^2(\mu_x) = \mu_x$
b) If $\nu = \overrightarrow{T}(\alpha|\mu_x)$ then $\Lambda_E^2(\nu) = \nu$.

Proof

a) Trivial (see Proposition 3.44)
b) According to Proposition 3.42 and Proposition 3.44,

$$\begin{aligned}
\Lambda_E^2(\nu) &= \Lambda_E(\Lambda_E(\overrightarrow{T}(\alpha|\mu_x))) \\
&\leq \Lambda_E(T(\alpha, \Lambda_E(\mu_x))) \\
&= \Lambda_E(T(\alpha, \mu_x)) = \overrightarrow{T}(\alpha|\Lambda_E(\mu_x)) \\
&= \overrightarrow{T}(\alpha|\mu_x) = \nu.
\end{aligned}$$

On the other hand, $\Lambda_E^2(\nu) \geq \nu$ (Corollary 3.50), so that $\Lambda_E^2(\nu) = \nu$.

Lemma 3.52. *Let $(\nu_i)_{i\in I}$ be a family of fixed points of Λ_E^2. Then $\bigwedge_{i\in I} \nu_i$ is also a fixed point of Λ_E^2.*

Proof. It follows from $\Lambda_E^2\left(\bigwedge_{i\in I} \nu_i\right) = \bigwedge_{i\in I}(\Lambda_E^2(\nu_i)) = \bigwedge_{i\in I} \nu_i$ (Proposition 3.48.a) and from $\Lambda_E^2(\bigwedge_{i\in I} \nu_i) \geq \bigwedge_{i\in I} \nu_i$ (Corollary 3.50).

Proposition 3.53. Fix(Λ_E^2) =Im(Λ_E).

Proof

a) $\mathrm{Im}(\Lambda_E) \subseteq \mathrm{Fix}(\Lambda_E^2)$:

$$\begin{aligned}\Lambda_E(\mu(x)) &= \inf_{y\in X} \overrightarrow{T}(\mu(y)|E(x,y)) \\ &= \inf_{y\in X} \overrightarrow{T}(\mu(y)|\mu_y(x)).\end{aligned}$$

For every $y \in X$, we can define a fuzzy subset ν_y in the following way: $\nu_y(x) = \overrightarrow{T}(\mu(y)|\mu_y(x))$ that is of the form of Lemma 3.51.b and therefore a fixed point of Λ_E^2. $\Lambda_E(\mu) = \inf_{y\in X} \nu_y$ which thanks to Lemma 3.52 belongs to $\mathrm{Fix}(\Lambda_E^2)$ as well. So, $\mathrm{Im}(\Lambda_E) \subseteq \mathrm{Fix}\ (\Lambda_E^2)$.

b) Fix $(\Lambda_E^2) \subseteq \mathrm{Im}(\Lambda_E)$: Given $\nu \in [0,1]^X$ such that $\Lambda_E^2(\nu) = \nu$, then $\nu \in \mathrm{Im}(\Lambda_E)$ because

$$\nu = \Lambda_E^2(\nu) = \Lambda_E(\Lambda_E(\nu)).$$

As a consequence of Proposition 3.53 we can easily check that $\mathrm{Im}(\Lambda_E) \subsetneq H_E$, as it is shown in the next example:

Example 3.54. Let be $X = \{x_1, x_2, x_3\}$, T the Łukasiewicz t-norm, E the T-indistinguishability operator defined by $E(x_i, x_j) = 0$ if $i \neq j$ and $\mu \in H_E$ defined by $\mu(x_1) = 1$, $\mu(x_2) = 0.5$, $\mu(x_3) = 0$.

We have that $\Lambda_E(\mu) = (0.5, 0, 0)$ and $\Lambda_E^2(\mu) = (1, 0.5, 0.5) \neq \mu = (1, 0.5, 0)$ and $\mu \notin \mathrm{Im}(\Lambda_E)$.

Corollary 3.55. $\Lambda_E^3 = \Lambda_E$.

Proof. Consequence of Proposition 3.53.

Corollary 3.56. $\Lambda_E^{2n} = \Lambda_E^2$, $\Lambda_E^{2n+1} = \Lambda_E$ *with* $n \in \mathbb{N}$.

In particular, Λ_E^2 is a fuzzy closure operator and $\mathrm{Im}\Lambda_E$ is the set of closed sets of a fuzzy topology.

In Proposition 3.43 we have proved that the set $\mathrm{Fix}(\Lambda_E)$ of fixed points of Λ_E is contained in the set P_X of fuzzy points of E. Now we will characterize the fixed points of Λ_E as exactly the maximal fuzzy points of E. Moreover, given a fuzzy point μ, we can find a fixed point μ' of Λ_E with $\mu \leq \mu'$.

Considering the natural T-indistinguishability operator E_T (cf. Definition 3.79), we have an isometric embedding of (X, E) into $(\mathrm{Fix}(\Lambda_E), E_T)$. Some of its properties will be shown.

Lemma 3.57. *Let E be a T-indistinguishability operator on X and $\mu \in H_E$. $\Lambda_E(\mu) \geq \mu$ if and only if $\mu \in P_X$.*

Proof

$$\begin{aligned} \Lambda_E(\mu)(x) = \inf_{y\in X} \overrightarrow{T}(\mu(y)|E(x,y)) \geq \mu(x) \\ \Leftrightarrow \overrightarrow{T}(\mu(y)|E(x,y)) \geq \mu(x)\ \forall x,y \in X \\ \Leftrightarrow T(\mu(x),\mu(y)) \leq E(x,y)\ \forall x,y \in X. \end{aligned}$$

The next Theorem characterizes the set of fixed points of Λ_E.

Theorem 3.58. *Let E be a T-indistinguishability operator on X.* Fix(Λ_E) *is the set of all fuzzy points $\mu \in P_X$ which are maximal in P_X.*

Proof

a) Let μ be a fixed point of Λ_E and $\mu' \in P_X$ with $\mu \leq \mu'$.

$$\mu(x) = \inf_{y\in Y} \overrightarrow{T}(\mu(y)|E(x,y)) \geq \inf_{y\in Y} \overrightarrow{T}(\mu'(y)|E(x,y)) \geq \mu'(x).$$

So, $\mu = \mu'$.

b) Let μ be a fuzzy point not in Fix(Λ_E). Then there exists $x_0 \in X$ with $\mu(x_0) < \inf_{y\in Y} \overrightarrow{T}(\mu(y), E(x_0,y))$.

We can define a new fuzzy subset μ' by

$$\mu'(x) = \begin{cases} \mu(x) & \text{if } x \neq x_0 \\ \inf_{y\in Y} \overrightarrow{T}(\mu(y), E(x_0,y)) & \text{otherwise.} \end{cases}$$

μ' is a fuzzy point and $\mu' > \mu$ which means that μ is not maximal in P_X.

Using Zorn's Lemma, we can see that every fuzzy point is contained in a fixed point of Λ_E.

Corollary 3.59. *Given a fuzzy point μ, there exists a fixed point μ' of Λ_E with $\mu \leq \mu'$.*

Proposition 3.60. *Let E be a T-indistinguishability operator on a finite set X and μ a non-normal fuzzy subset of X. $\Lambda_E(\mu) = \mu$ if and only if $\mu = \mu_a$ ($a \notin X$) satisfying $\forall x \in X\ \exists u_x \in X$ such that $\overrightarrow{T}\left(\overline{E}(a,u_x)|E(x,u_x)\right) = \overline{E}(x,a)$.*

Proof

$\Rightarrow$) Let us suppose that $\Lambda_E(\mu) = \mu$. In this case, $\mu \in C_E$ (Proposition 3.43) and being $\mu(x) < 1$ for all $x \in X$, we have that $\mu = \mu_a$, $a \notin X$. Further

$$\Lambda_E(\mu_a)(x) = \inf_{y\in X} \overrightarrow{T}(\mu_a(y)|E(y,x)) = \mu_a(x)$$

which, being X finite, implies that for all $x \in X$ there exists u_x such that

$$\begin{aligned}\overrightarrow{T}(\mu_a(u_x)|E(u_x,x)) &= \overrightarrow{T}(\overline{E}(a,u_x)|E(x,u_x))\\ &= \mu_a(x) = \overline{E}(x,a).\end{aligned}$$

$\Leftarrow$) Let $\mu = \mu_a$ ($a \notin X$) be a fuzzy subset satisfying that for all x there exists u_x such that $\overrightarrow{T}(\overline{E}(a,u_x)|E(x,u_x)) = \overline{E}(x,a)$. In this case

$$\begin{aligned}\Lambda_E(\mu)(x) &= \inf_{y\in X} \overrightarrow{T}(\mu(y)|E(y,x))\\ &\geq \inf_{y\in X\cup\{a\}} \overrightarrow{T}(\mu(y)|\overline{E}(y,x)) \geq \mu(x)\end{aligned}$$

for all $x \in X$. On the other hand

$$\begin{aligned}\inf_{y\in X} \overrightarrow{T}(\mu(y)|E(y,x)) &\leq \overrightarrow{T}(\mu(u_x)|E(x,u_x))\\ &= \overrightarrow{T}(\overline{E}(a,u_x)|E(x,u_x))\\ &= E(x,a) = \mu(x),\end{aligned}$$

so that $\Lambda_E(\mu)(x) = \mu(x)$ for all $x \in X$.

This proposition can be easily extended to non-finite sets X by replacing the condition $\forall x \in X\ \exists u_x \in X$ such that $\overrightarrow{T}(\overline{E}(a,u_x)|E(x,u_x)) = E(x,u_x)$ by the more technical one

$$\forall x \in X\ \forall \epsilon \in [0,1]\ \exists u_{x,\epsilon} \text{ such that } \overrightarrow{T}(\overline{E}(a,u_{x,\epsilon})|E(x,u_{x,\epsilon})) < E(x,a) + \epsilon.$$

The proof is similar to that of Proposition 3.44.

There is a nice relation between the couples of fuzzy subsets μ, μ' of $\mathrm{Im}(\Lambda_E)$ which are one image of the other one that will be studied next. We shall call μ and μ' dual fuzzy subsets.

Proposition 3.61. *Let μ be a fixed point of Λ_E and $\alpha \in [0,1]$. If $T(\alpha,\mu)$ and $\overrightarrow{T}(\alpha|\mu)$ are in* $\mathrm{Im}(\Lambda_E)$*, then they are dual fuzzy subsets.*

Proof. It is a consequence of Proposition 3.42:

3.42 a) states that

$$\Lambda_E(T(\alpha,\mu)) = \overrightarrow{T}(\alpha|\mu)).$$

On the other hand,

$$\Lambda_E(\overrightarrow{T}(\alpha|\mu) = \Lambda_E(\overrightarrow{T}(\alpha|\Lambda_E(\mu)) = \Lambda_E^2(T(\alpha,\mu)) = T(\alpha,\mu),$$

where the last equality follows from Proposition 3.53.

If T is a continuous Archimedean t-norm with additive generator t, then we can associate to T the quasi-arithmetic mean m_t generated by t, i.e. $m(x,y) = t^{-1}\left(\frac{t(x)+t(y)}{2}\right)$ for all $x,y \in [0,1]$ (cf. Definition 2.64). Then a fixed point μ of Λ happens to be the quasi-arithmetic mean of the dual fuzzy subsets $T(\alpha,\mu)$ and $\overrightarrow{T}(\alpha|\mu)$.

Proposition 3.62. *Let μ be a fixed point of Λ_E, $\alpha \in [0,1]$ and $T(\alpha,\mu)$ and $\overrightarrow{T}(\alpha|\mu)$ dual non-normalized fuzzy subsets in* $\mathrm{Im}(\Lambda_E)$ *with T a continuous Archimedean t-norm with additive generator t. Then μ is the quasi-arithmetic mean of these dual fuzzy subsets.*

Proof

$$m_t(T(\alpha,\mu), \overrightarrow{T}(\alpha|\mu)) = t^{-1}\left(\frac{T(\alpha,\mu) + \overrightarrow{T}(\alpha|\mu)}{2}\right)$$

$$= t^{-1}\left(\frac{t\left(t^{[-1]}(t(\alpha)+t(\mu))\right) + t\left(t^{[-1]}(t(\mu)-t(\alpha))\right)}{2}\right)$$

$$= t^{-1}\left(\frac{(t(\alpha)+t(\mu)+t(\mu)-t(\alpha)}{2}\right) = \mu.$$

This means in particular that these dual fuzzy subsets and μ generate the same T-indistinguishability operator and the same T-preorder.

The fuzzy relation E_T on $[0,1]^X$ defined for all $\mu, \mu' \in [0,1]^X$ by

$$E_T(\mu,\mu') = \inf_{x\in X} E_T(\mu(x), \mu'(x))$$

is a very important T-indistinguishability operator called the natural T-indistinguishability operator on $[0,1]^X$ (cf. Section 3.4).

Restricting E_T to the set P_X of fuzzy points of X, we have the following result.

Proposition 3.63. *Let E be a T-indistinguishability operator on X. If μ is a fixed point of Λ_E and μ_x is the column corresponding to the element x of X, then*

$$E_T(\mu,\mu_x) = \mu(x).$$

Proof

$$\begin{aligned}
\mu(y) &= \inf_{z\in X} \overrightarrow{T}(\mu(z)|E(y,z)) \\
&\geq \inf_{z\in X} \overrightarrow{T}(\mu(z)|T(E(y,x),E(x,z))) \\
&\geq_{(*)} T(E(y,x), \inf_{z\in X} \overrightarrow{T}(\mu(z)|E(x,z))) \\
&= T(\mu_x(y), \mu(x))
\end{aligned}$$

where $(*)$ follows from Lemma 2.40.

Therefore

$$\mu(x) \leq \overrightarrow{T}(\mu_x(y)|\mu(y)).$$

On the other hand, since μ is a fuzzy point,

$$\mu_x(y) = E(x,y) \geq T(\mu(x), \mu(y))$$

or, equivalently,

$$\mu(x) \leq \overrightarrow{T}(\mu(y)|\mu_x(y)).$$

$$\mu(x) \leq \min(\overrightarrow{T}(\mu_x(y)|\mu(y)), \overrightarrow{T}(\mu(y)|\mu_x(y))) \ \forall x, y \in X$$

and therefore

$$\mu(x) \leq \inf_{y \in X} \min(\overrightarrow{T}(\mu_x(y)|\mu(y)), \overrightarrow{T}(\mu(y)|\mu_x(y))) = E_T(\mu_x, \mu).$$

But since

$$\min(\overrightarrow{T}(\mu_x(x)|\mu(x)), \overrightarrow{T}(\mu(x)|\mu_x(x))) = \mu(x),$$

we finally get our result.

Corollary 3.64. *Let E be a T-indistinguishability operator on X. The map $\tau : X \to \mathrm{Fix}(\Lambda_E)$ defined by $\tau(x) = \mu_x$ is an isometric embedding.*

Proof. Trivial: $E_T(\mu_x, \mu_y) = \mu_y(x) = \mu_x(y) = E(x, y)$.

Corollary 3.65. *Let E be a T-indistinguishability operator on X and μ, μ' fixed points of Λ_E. Then*

$$E_T(\mu, \mu') \geq T(\mu(x), \mu'(x)) \ \forall x \in X$$

Proof

$$E_T(\mu, \mu') \geq T(E_T(\mu, \mu_x), E_T(\mu_x, \mu')) = T(\mu(x), \mu'(x)).$$

Proposition 3.66. *Let E be a T-indistinguishability operator on X and $\mu, \mu' \in P_X$. Then*

$$E_T(\mu, \mu') \leq E_T(\Lambda_E(\mu), \Lambda_E(\mu')).$$

Proof

$$\begin{aligned}
\Lambda_E(\mu)(x) &= \inf_{y \in X} \overrightarrow{T}(\mu(y)|E(x, y)) \\
&\geq \inf_{y \in X} T(\overrightarrow{T}(\mu(y))|\mu'(y), \overrightarrow{T}(\mu'(y)|E(x, y))) \\
&\geq T(\inf_{y \in X} \overrightarrow{T}(\mu(y)|\mu'(y)), \inf_{y \in X} \overrightarrow{T}(\mu'(y)|E(x, y))) \\
&\geq T(\inf_{y \in X} \overrightarrow{T}(\mu(y)|\mu'(y)), \Lambda_E(\mu')(x))
\end{aligned}$$

and therefore,

$$\overrightarrow{T}(\Lambda_E(\mu')(x)|\Lambda_E(\mu)(x)) \geq \inf_{y \in X} \overrightarrow{T}(\mu(y)|\mu'(y)) \geq E_T(\mu, \mu').$$

Similarly,

$$\overrightarrow{T}(\Lambda_E(\mu)(x)|\Lambda_E(\mu')(x)) \geq \inf_{y \in X} \overrightarrow{T}(\mu'(y)|\mu(y)) \geq E_T(\mu, \mu')$$

and

$$E_T(\Lambda_E(\mu), \Lambda_E(\mu')) \geq E_T(\mu, \mu').$$

Let us conclude this section with a very simple example that gives a geometrical interpretation of the maps and sets studied so far in this chapter.

Example 3.67. Let $X = \{a, b\}$ and consider the T-indistinguishability operator E with $E(a, b) = m$. Every fuzzy subset μ of X can be identified with the point $(\mu(a), \mu(b))$ of $[0, 1]^2$.

H_E, the set of extensional sets of E is then the region of $[0, 1]^2$ defined by the inequation

$$E_T(x, y) \geq m$$

and P_X is the part of H_E limited by the inequation

$$T(x, y) \leq m.$$

If $\mu = (p, q)$, then $\Lambda_E(h) = (\overrightarrow{T}(q|m), \overrightarrow{T}(p|m))$ and $\Lambda_E^2(\mu) = (\overrightarrow{T}(\overrightarrow{T}(p|m)|m), \overrightarrow{T}(\overrightarrow{T}(q|m)|m))$.

If $\mu = (p, q)$ is not in H_E and $p > q$, then $\phi_E(\mu) = (p, T(m, p))$ and $\psi_E(\mu) = (\overrightarrow{T}(q|m), \overrightarrow{T}(p|m))$. If $p < q$, then $\phi_E(\mu) = (T(m, q), q)$ and $\psi_E(\mu) = (\overrightarrow{T}(q|m), \overrightarrow{T}(p|m))$.

Taking $m = 0.4$ and T the product t-norm, H_E is the region of $[0, 1]^2$ defined by the inequations

$$x - 0.4y \geq 0$$
$$0,4x - y \geq 0$$

and P_X is the part of this region below the hyperbola

$$xy = 0.4.$$

The fixed points of Λ_E are the maximal elements of P_X and therefore are the points in this hyperbola.

If $\mu = (p, q)$, then $\Lambda_E(\mu) = (\min(1, \frac{0.4}{q}), \min(1, \frac{0.4}{p}))$ and $\Lambda_E^2(\mu) = (\max(p, 0.4), \max(q, 0.4))$. $\text{Fix}(\Lambda_E^2) = \text{Im}(\Lambda_E)$ is the square

$$0.4 \leq x \leq 1$$
$$0.4 \leq y \leq 1.$$

In this set, the image under Λ_E of a fuzzy subset bellow the hyperbola $xy = 0.4$ (i.e.: bellow $\text{Fix}(\Lambda_E)$) is a point above it and vice versa, the hyperbola being a kind of symmetry axis, which gives a clear picture of corollary 3.56.

Finally, if $\mu = (p, q)$ is not in H_E and $p > q$,then $\phi_E(\mu) = (p, mp)$ and $\psi_E(\mu) = (\frac{m}{q}, q)$; if $p < q$, then $\phi_E(\mu) = (mq, q)$ and $\psi_E(\mu) = (p, \frac{m}{p})$. For example, if $\mu = (0.1, 0.8)$, $\phi_E(\mu) = (0.32, 0.8)$ and $\psi_E(\mu) = (0.1, 0.25)$ are

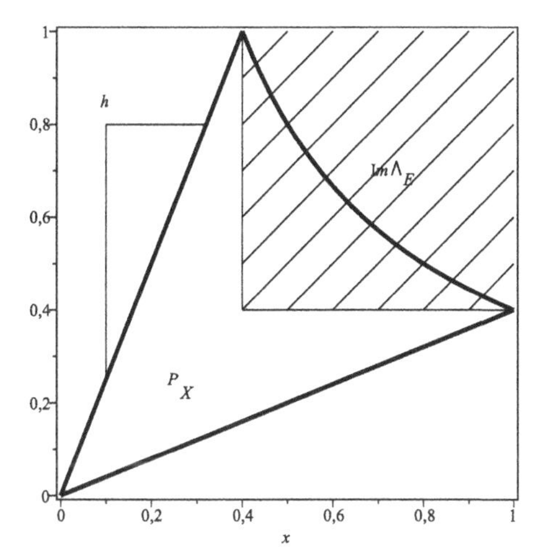

Fig. 3.1 The effect of ϕ_E, ψ_E and Λ_E

obtained by projecting μ to its closest edge of H_E horizontally and vertically respectively. See Figure 3.1.

3.3 Fuzzy Points and the Representation Theorem

In this section we are interested in finding when a family $(\mu_i)_{i\in I}$ of fuzzy subsets of X are columns of the T-indistinguishability operator they generate.

The first result tells that the T-indistinguishability operator on X generated by a family $(\mu_i)_{i\in I}$ of fuzzy subsets of X is the greatest (less specific) of such operators for which all the elements of the family are extensional.

Proposition 3.68. *Let $F = (\mu_i)_{i\in I}$ be a family of fuzzy subsets of X and E the T-indistinguishability operator generated by this family ($E(x,y) = \inf_{i\in I} E_{\mu_i}(x,y)$). Then E is the greatest T-indistinguishability operator for which all the fuzzy subsets of the family are extensional.*

Proof. We already know that, since the elements of F are generators of E, they also are extensional.

Let us prove that E is the greatest T-indistinguishability operators satisfying this property.

Let E' be another T-indistinguishability operator with all the fuzzy subsets of F extensional. Then $E' \leq E_{\mu_i}$ for all $i \in I$ and therefore $E' \leq \inf_{i\in I} E_{\mu_i} = E$.

Proposition 3.69. *Let $(\mu_i)_{i\in I}$ be a family of normal fuzzy subsets of X and $(x_i)_{i\in I}$ a family of elements of X such that $\mu_i(x_i) = 1$ for all $i \in I$. Then the following two properties are equivalent.*

a) There exists a T-indistinguishability operator E on X such that

$$\mu_i(x) = E(x, x_i) \ \forall i \in I \ \forall x \in X.$$

b) For all $i, j \in I$,

$$\sup_{x \in X} T(\mu_i(x), \mu_j(x)) \leq \inf_{y \in X} E_T(\mu_i(y), \mu_j(y)).$$

Proof

a) $\Rightarrow$ b) It suffices to prove

$$T(\mu_i(x), \mu_j(x)) \leq E_T(\mu_i(y), \mu_j(y))$$

for all $i \in I$, which is equivalent to prove

$$T(E(x, x_i), E(x, x_j)) \leq E_T(E(y, x_i), E(y, x_j)).$$

But from 2.51 we have

$$T(E(x, x_i), E(x, x_j)) \leq E(x_i, x_j) \leq E_T(E(y, x_i), E(y, x_j)).$$

b) $\Leftarrow$ a)

$$E(x, x_i) \leq E_{\mu_i}(x, x_i) = E_T(\mu_i(x), \mu_i(x_i)) = E_T(\mu_i(x), 1) = \mu_i(x).$$

We need to prove the other inequality $\mu_i \leq E(x, x_i)$, which is equivalent to prove

$$\mu_i(x) \leq E_{\mu_j}(x, x_i) = \min(\overrightarrow{T}(\mu_j(x)|\mu_j(x_i)), \overrightarrow{T}(\mu_j(x_i)|\mu_j(x))) \ \forall j \in I.$$

(i) To prove $\mu_i(x) \leq \overrightarrow{T}(\mu_j(x)|\mu_j(x_i))$ is equivalent to prove

$$T(\mu_i(x), \mu_j(x)) \leq \mu_j(x_i):$$

$$\begin{aligned} T(\mu_i(x), \mu_j(x)) &\leq \sup_{z \in X} T(\mu_i(z), \mu_j(z)) \\ &\leq \inf_{y \in X} E_T(\mu_i(y), \mu_j(y)) \\ &\leq E_T(\mu_i(x_i), \mu_j(x_i)) \\ &= E_T(1, \mu_j(x_i)) = \mu_j(x_i). \end{aligned}$$

(ii)To prove $\mu_i(x) \leq \overrightarrow{T}(\mu_j(x_i)|\mu_j(x))$ is equivalent to prove

$$T(\mu_i(x), \mu_j(x_i)) \leq \mu_j(x)$$

or equivalently

$$\mu_j(x_i) \leq \overrightarrow{T}(\mu_i(x)|\mu_j(x)):$$

$$\begin{aligned}\overrightarrow{T}(\mu_i(x)|\mu_j(x)) &\geq E_T(\mu_i(x), \mu_j(x)) \\ &\geq \inf_{y\in X} E_T(\mu_i(y), \mu_j(y)) \\ &\geq \sup_{z\in X} T(\mu_i(z), \mu_j(z)) \\ &\geq T(\mu_i(x_i), \mu_j(x_i)) \\ &= T(1, \mu_j(x_i)) = \mu_j(x_i).\end{aligned}$$

Corollary 3.70. *Let $(\mu_i)_{i\in I}$ be a family of normal fuzzy subsets of X and $(x_i)_{i\in I}$ a family of elements of X such that $\mu_i(x_i) = 1$ for all $i \in I$ satisfying*

$$\sup_{x\in X} T(\mu_i(x), \mu_j(x)) \leq \inf_{y\in X} E_T(\mu_i(y), \mu_j(y)).$$

for all $i, j \in I$. $E = \inf_{i\in I} E_{\mu_i}$ is the greatest T-indistinguishability operator satisfying

$$\mu_i(x) = E(x, x_i)\ \forall i \in I,\ \forall x \in X.$$

Proof. The sets μ_i are extensional with respect to E. If a T-indistinguishability E' has the sets μ_i as columns, then they are also extensional with respect to E' because

$$T(E'(x,y), \mu_i(x)) = T(E'(x,y), E'(x,x_i) \leq E'(y,x_i)) = \mu_i(y)$$

and from Proposition 3.68 the result follows.

Lemma 3.71. *Let $(\mu_i)_{i\in I}$ be a family of normal fuzzy subsets of X and $(x_i)_{i\in I}$ a family of elements of X such that $\mu_i(x_i) = 1$ for all $i \in I$ satisfying*

$$\sup_{x\in X} T(\mu_i(x), \mu_j(x)) \leq \inf_{y\in X} E_T(\mu_i(y), \mu_j(y)).$$

for all $i, j \in I$. Then $E = \sup_{i\in I} E^{\mu_i}$ is a T-indistinguishability operator on X, where E^{μ_i} is the decomposable T-indistinguishabilty operator generated by μ_i.

Proof. E is clearly reflexive and symmetric. Let us prove that it is also T-transitive.

$$\begin{aligned}T(E(x,y), E(y,z)) &= \sup_{i\in I, j\in I} T(\mu_i(x), \mu_i(y), \mu_j(y), \mu_j(z)) \\ &\leq \sup_{i\in I, j\in I} T(\mu_i(x), (\sup_{s\in X} T(\mu_i(s), \mu_j(s))), \mu_j(z)) \\ &\leq \sup_{i\in I, j\in I} T(\mu_i(x), (\inf_{s\in X} E_T(\mu_i(s), \mu_j(s))), \mu_j(z)) \\ &\leq \sup_{i\in I, j\in I} T(\mu_i(x), E_T(\mu_i(x), \mu_j(x)), \mu_j(z)) \\ &\leq \sup_{i\in I, j\in I} T(\mu_i(x), \overrightarrow{T}(\mu_i(x)|\mu_j(x)), \mu_j(z))\end{aligned}$$

$$\leq \sup_{i\in I, j\in I} T(\mu_j(x), \mu_j(z))$$
$$\leq \sup_{j\in I} T(\mu_j(x), \mu_j(z)) = E(x, z).$$

Proposition 3.72. *Let $(\mu_i)_{i\in I}$ be a family of normal fuzzy subsets of X and $(x_i)_{i\in I}$ a family of elements of X such that $\mu_i(x_i) = 1$ for all $i \in I$ satisfying*

$$\sup_{x\in X} T(\mu_i(x), \mu_j(x)) \leq \inf_{y\in X} E_T(\mu_i(y), \mu_j(y)).$$

for all $i, j \in I$. Then the T-indistinguishability operator $E = \sup_{i\in I} E^{\mu_i}$ is the smallest T-indistinguishability operator on X satisfying

$$\mu_i(x) = E(x, x_i)\ \forall i \in I,\ \forall x \in X.$$

Proof. We will first prove that $E(x, x_j) = \mu_j(x)$.

$$E(x, x_j) = \sup_{i\in I} T(\mu_i(x), \mu_i(x_j)) \geq T(\mu_j(x), \mu_j(x_j)) = \mu_j(x).$$

Let us prove the other inequality.

$$\begin{aligned}
E(x, x_j) &= T(\mu_j(x_j), E(x, x_j)) \\
&= \sup_{i\in I} T(\mu_j(x_j), \mu_i(x), \mu_i(x_j)) \\
&\leq \sup_{i\in I} T(\left(\sup_{y\in X} T(\mu_j(y), \mu_i(y))), \mu_i(x)\right) \\
&\leq \sup_{i\in I} T(\left(\inf_{y\in X} E_T(\mu_j(y), \mu_i(y))), \mu_i(x)\right) \\
&\leq \sup_{i\in I} T(E_T(\mu_j(x), \mu_i(x)), \mu_i(x)) \\
&\leq \sup_{i\in I} T(\overrightarrow{T}(\mu_i(x)|\mu_j(x)), \mu_i(x)) \\
&\leq \mu_j(x).
\end{aligned}$$

Let us prove now the minimality of E.

If E' satisfies $E'(x, x_j) = \mu_j(x)$, then

$$E'(x, y) \geq T(E(x, x_i), E(x_i, y)) = T(\mu_i(x), \mu_i(y))\ \forall i \in I.$$

Therefore $E'(x, y) \geq \sup_{i\in I} T(\mu_i(x), \mu_i(y)) = E(x, y)$.

3.4 Indistinguishability Operators Between Fuzzy Subsets: Duality Principle

In many situations there is a need of measuring the similarity or indistinguishability between fuzzy subsets of a universe of discourse X. This can be done generalizing the natural indistinguishability operator E_T. Also if there is a T-indistinguishability operator defined on X, it can be useful to extend it to the fuzzy subsets of X. In this section, these generalizations will be done in the framework of the duality principle [108],[15].

Fuzzy set theory has a kind of asymmetry in the sense that while the fuzzy subsets are vague, it is not the case with the elements of X. With the duality principle, the elements of X can be seen as fuzzy sets acting on $[0,1]^X$ and this asymmetry can be overcome in some sense.

Definition 3.73. *Let H be a family of fuzzy subsets of X. H separates points if and only if for any $x, y \in X$ with $x \neq y$, there exists $\mu \in H$ such that $\mu(x) \neq \mu(y)$.*

Definition 3.74. *Let H be a family of fuzzy subsets of X. The map $B_H : X \to H$ is defined for all $x \in X$ by*

$$B_H(x) = \mu(x).$$

$B_H(x)$ will also be indicated by $= x^{**}(\mu)$.

Proposition 3.75. *B_H is a one to one map if and only if H separates points.*

Proof. $\Rightarrow$) If B_H is one to one, then for $x \neq y$ we have $x^{**} \neq y^{**}$. Therefore there exists $\mu \in H$ such that $x^{**}(\mu) \neq y^{**}(\mu)$ and hence $\mu(x) \neq \mu(y)$.

$\Leftarrow$) If H separates points and $x^{**} = y^{**}$, then $x^{**}(\mu) = y^{**}(\mu)$ for all $\mu \in H$ and therefore $x = y$.

Let us now introduce indistinguishability operators on the universes of discourse.

Definition 3.76. *Let E, F be T-indistinguishability operators on X and Y respectively. $\varphi : X \to Y$ is a homomorphism if and only if*

$$F(\varphi(x_1), \varphi(x_2)) = E(x_1, x_2)$$

for all $x_1, x_2 \in X$.

If φ is a one to one map, then it is called a monomorphism and if in addition it is onto, an isomorphism.

Definition 3.77. *Let E, F be T-indistinguishability operators on X and Y respectively. Y is an extension of X if and only if there exists a one to one homomorphism $\varphi : X \to Y$ (cf. Definition 3.37).*

Definition 3.78. *Let E, F be two T-indistinguishability operators on X and Y respectively. $\varphi : X \to Y$ is extensional if and only if*

$$F(\varphi(x_1), \varphi(x_2)) \geq E(x_1, x_2)$$

for all $x_1, x_2 \in X$.

In this context, extensional fuzzy subsets with respect to a T-indistinguishability operator E are exactly the extensional maps $\varphi : X \to [0,1]$, where we consider the natural T-indistinguishability operator E_T on $[0,1]$.

Let us now extend the natural T-indistinguishability operator E_T to the set of fuzzy subsets.

Definition 3.79. *Let X be a set and $H \subseteq [0,1]^X$ a set of fuzzy subsets of X. The natural T-indistinguishability operator on H is defined for all $\mu, \nu \in H$ by*

$$E_T(\mu, \nu) = \inf_{x \in X} E_T(\mu(x), \nu(x)).$$

Proposition 3.80. *The natural T-indistinguishability operator on H is indeed a T-indistinguishability operator.*

Proof

$$E_T(\mu, \nu) = \inf_{x \in X} E_T(\mu(x), \nu(x)) = \inf_{x^{**} \in B_H(X)} E_T(x^{**}(\mu), x^{**}(\nu)).$$

Since x^{**} are fuzzy subsets of H, E_T is a T-indistinguishability operators by the Representation Theorem.

The natural indistinguishability operator E_T between fuzzy subsets is widely used to compare them. To decide the degree in which two fuzzy subsets μ and ν are indistinguishables we first compare all their images $\mu(x)$ and $\nu(x)$ with the natural T-indistinguishability operator E_T $(E_T(\mu(x), \nu(x))$ and then the infimum of these values are taken. From a logical point of view, this fuzzify the sentence

> μ and ν are equivalent or indistinguishable if and only if for all $x \in X$ $\mu(x)$ and $\nu(x)$ are.

Though it is the most natural way to compare fuzzy subsets, the use of the infimum is very drastic and can produce dramatic effects. Let us suppose for instance that μ and ν coincide in all points but one x_0 and for that particular one $\mu(x_0) = 1$ and $\nu(x_0) = 0$. then $E_T(\mu, \nu) = 0$ which would not be very reasonable in many real situations. In Chapter 8 some alternatives are proposed.

Let us go back to the map ϕ_E.

Proposition 3.81. *Let E_T be the natural T-indistinguishability operator on $[0,1]^X$ and E a T-indistinguishability operator on X separating points. Then $\phi_E : [0,1]^X \to [0,1]^X$ is extensional.*

Proof. We must prove that $E_T(\phi_E(\mu), \phi_E(\nu)) \geq E_T(\mu, \nu)$ for all $\mu, \nu \in [0,1]^X$.

Fixing $x \in X$,

$$\begin{aligned}
&\overrightarrow{T}(\phi_E(\mu)(x)|\phi_E(\nu)(x)) \\
&= \overrightarrow{T}(\sup_{y\in X} T(E(x,y),\mu(y))| \sup_{y\in X} T(E(x,y),\nu(y))) \\
&= \inf_{z\in X} \overrightarrow{T}(T(E(x,y),\mu(y))| \sup_{y\in X} T(E(x,y),\nu(y))) \\
&\geq \inf_{z\in X} \overrightarrow{T}(T(E(x,y),\mu(y))|T(E(x,y),\nu(y))) \\
&\geq_{(*)} \inf_{z\in X} \overrightarrow{T}(\mu(z)|\nu(z)) \\
&\geq \inf_{z\in X} E_T(\mu(z),\nu(z)) = E_T(\mu,\nu)
\end{aligned}$$

where $(*)$ follows from Lemma 2.43.

Similarly we can obtain

$$\overrightarrow{T}(\phi_E(\nu)(x)|\phi_E(\mu)(x)) \geq E_T(\mu,\nu)$$

and therefore

$$E_T(\phi_E(\mu)(x),\phi_E(\nu)(x)) \geq E_T(\mu,\nu).$$

Finally

$$\inf_{x\in X} E_T(\phi_E(\mu)(x),\phi_E(\nu)(x)) = E_T(\phi_E(\mu),\phi_E(\nu)) \geq E_T(\mu,\nu).$$

Proposition 3.82. *Let E be a T-indistinguishability operator on a set X and $x,y\in X$. Then $E(x,y) = E_T(\phi_E(\{x\}),\phi_E(\{y\})) = E_T(\mu_x,\mu_y)$ where μ_x and μ_y are the columns of E associated to x and y respectively.*

Proof

$$\begin{aligned}
E_T(\mu_x,\mu_y) &= \inf_{z\in X} E_T(\mu_x(z),\mu_y(z)) \\
&= \inf_{z\in X} E_T(E(x,z),E(y,z)) \\
&= \inf_{z\in X} E_T(\mu_z(x),\mu_z(y)) \\
&= \inf_{z\in X} E_{\mu_z}(x,y) =_{(*)} E(x,y).
\end{aligned}$$

(*) follows because the columns of E are a generating family of E.

This proposition says that $\phi_E : X \to H_E$ (E_T in H_E) is a monomorphism for any T-indistinguishability E.

If there is a T-indistinguishability E on X, then the most natural way to extend it to fuzzy sets is inspired in the last result.

Definition 3.83. *Let E be a T-indistinguishability operators on a set X. E is extended to $[0,1]^X$ by*

$$E(\mu,\nu) = E_T(\phi_E(\mu),\phi_E(\nu)).$$

for all fuzzy subsets μ,ν of X.

Definition 3.84. *The dual map of* $\varphi : X \to Y$ *is defined by*

$$\begin{aligned} \varphi^T : [0,1]^Y &\to [0,1]^X \\ \mu &\to \varphi^T(\mu) = \mu \circ \varphi. \end{aligned}$$

Lemma 3.85. *Let* E *and* F *be* T*-indistinguishability operators on* X *and* Y *respectively. If* $\varphi : X \to Y$ *is extensional, then* $\varphi^T : H_F \to H_E$ *is extensional.*

Proof. If φ is extensional, then

$$\begin{aligned} E_T(\varphi^T(\nu_1), \varphi^T(\nu_2)) &= \inf_{x \in X} E_T(\varphi^T(\nu_1)(x), \varphi^T(\nu_2)(x)) \\ &= \inf_{x \in X} E_T(\nu_1 \circ \varphi(x), \nu_2 \circ \varphi(x)) \\ &\geq \inf_{y \in Y} E_T(\nu_1(y), \nu_2(y)) = E_T(\nu_1, \nu_2). \end{aligned}$$

In fact, the reciprocal of the lemma is also true [13].

Proposition 3.86. *Let* E *and* F *be* T*-indistinguishability operators on* X *and* Y *respectively and* $\varphi : X \to Y$ *a homomorphism. Then there exists* $f : Y \to H_E$ *such that*

1. $f(Y) \subseteq H_E$ *is a generating system of* E.
2. f *is extensional with respect to* F *and* E_T.
3. $f(y_1) = f(y_2)$ *if and only if* $F(y_1, y) = F(y_2, y)$ *for all* $y \in \varphi(X)$.

Proof. Let us consider $f = \varphi^T \circ \phi_F$ (i.e. $f(y) = \varphi^T \circ \phi_F(y) = \phi_F(y) \circ \varphi$, i.e. $f(y) = F(y, \varphi(x))$ for all $x \in X$).

1.

$$\begin{aligned} &\inf_{y \in Y} E_T(f(y)(x_1), f(y)(x_2)) \\ &= \inf_{y \in Y} E_T(F(y, \varphi(x_1)), F(y, \varphi(x_2))) \\ &= F(\varphi(x_1), \varphi(x_2)) = E(x_1, x_2). \end{aligned}$$

2. ϕ_F is a homomorphism and φ^T is extensional. Therefore $f = \varphi^T \circ f$ is extensional.
3. $\Rightarrow$) If $f(y_1) = f(y_2)$, then for all $x \in X$

$$F(y_1, \varphi(x)) = F(y_2, \varphi(x))$$

and then

$$F_{|\varphi(X) \times \varphi(X)}(y_1, y_2) = \inf_{x \in X} E_T(F(\varphi(x), y_1), F(\varphi(x), y_2)) = 1.$$

$\Leftarrow$) If $F_{|\varphi(X) \times \varphi(X)}(y_1, y_2) = 1$, the only possibility is $F(\varphi(x), y_1) = F(\varphi(x), y_2)$ for all $x \in X$ and hence $f(y_1) = f(y_2)$.

Corollary 3.87. *In the last proposition f is a homomorphism if and only if $\phi_F(\varphi(X)) \subseteq H_F$ is a generating family of F.*

Proof

$$\begin{aligned} E_T(f(y_1), f(y_2)) &= \inf_{x \in X} E_T(f(y_1)(x), f(y_2)(x)) \\ &= \inf_{x \in X} E_T(F(y_1, \varphi(x)), F(y_2, \varphi(x))) \\ &= F(y_1, y_2) \end{aligned}$$

for all $y_1, y_2 \in Y$.

Let E be a T-indistinguishability operator on X and $\varphi : X \to X$ an isomorphism. We can consider the map

$$\begin{aligned} \overline{\varphi} : \phi_E(X) &\to \phi_E(X) \\ \phi_E(x) &\to \overline{\varphi}(\phi_E(x)) \end{aligned}$$

where

$$\begin{aligned} \overline{\varphi}(\phi_E(x)) : X &\to [0, 1] \\ u &\to \overline{\varphi}(\phi_E(x))(u) = \phi_E(\varphi(x))(u). \end{aligned}$$

Since ϕ_E is an isomomorphism, every element x of X can be identified with its image $\phi_E\ \{x\}) = E(x, \cdot)$ and we can identify $\overline{\varphi} = \varphi$.

Also φ^T is an isomorphism and we have the following commutative diagram.

$$\begin{array}{ccc} H_E & \xleftarrow{\varphi^T} & H_E \\ \uparrow{\scriptstyle \phi_E} & & \uparrow{\scriptstyle \phi_E} \\ X & \xrightarrow{\varphi} & X \\ \downarrow{\scriptstyle \phi_E} & & \downarrow{\scriptstyle \phi_E} \\ H_E & \xrightarrow{\overline{\varphi}} & H_E \end{array}$$

The next proposition shows the relation between φ and φ^T.

Proposition 3.88. *With the preceding notations, $\varphi^T = \varphi^{-1}$.*

Proof. Since φ^T is a bijective map, we must see that

$$\varphi^T(\phi_E(\varphi(x))) = \phi_E(x)$$

for all $x \in X$.

For $u \in X$ we have

$$\begin{aligned}\varphi^T(\phi_E(\varphi(x)))(u) &= \phi_E(\varphi(x))(\varphi(u)) \\ &= E(\varphi(x), \varphi(y)) \\ &= E(x, u) = \phi_E(x)(u).\end{aligned}$$

The fact that $\varphi^T = \varphi^{-1}$ resembles a similar situation of orthogonal endomorphisms in Vector Spaces. Indeed, let us suppose that $X = \{x_1, x_2, ..., x_n\}$ and $H = \{\mu_1, \mu_2, ..., \mu_m\}$ are finite sets of cardinality n and m respectively. $\varphi : H \to [0,1]^X$ can be represented by a matrix $M = (m_{ij})$, $i = 1, 2, ..., n$, $j = 1, 2, ..., m$ with $m_{ij} = \varphi(\mu_j)(x_i)$. Then the matrix of $\varphi^{-1} = \varphi^T$ is M^T, the transposed matrix of M. Indeed,

$$m_{ij} = \varphi(\phi_E(x_j))(x_i) = \phi_E(\varphi(x_j))(x_i) = E(x_i, \varphi(x_j))$$

and if $N = (n_{ij})$ is the matrix of $\varphi^{-1} = \varphi^T$,

$$n_{ij} = \varphi^T(\phi_E(x_j))(x_i) = \phi_E(x_j)(\varphi(x_i)) = E(\varphi(x_i), x_j).$$

Therefore $N = M^T$.

Let us end this chapter with another characterization of the fuzzy points of a T-indistinguishability operator.

Given a T-indistinguishability operator E on a set X, we can consider the set H_E of extensional fuzzy subsets with respect to E, the natural T-indistinguishability operator E_T on H_E and the operator $\phi_{E_T} : [0,1]^{H_E} \to [0,1]^{H_E}$. The composition

$$\phi_{E_T} \circ \phi_E : X \to H_{E_T} \subseteq [0,1]^{H_E}$$

is a monomorphism.

Theorem 3.89. *With the previous notations, $\mu \in H_E$ is a fuzzy point if and only if*

$$\phi_{E_T} \circ \phi_E(x)(\mu) = x^{**}(\mu).$$

Proof. $\Rightarrow$) If μ is a fuzzy point, then there exists an extension (Y, F) with $Y = X \cup \{y\}$ of (X, E) with $\mu = F(y, \cdot)$.

In general

$$\begin{aligned}\phi_{E_T} \circ \phi_E(x)(\mu) &= \phi_{E_T}(\phi_E(x))(\mu) \\ &= E_T(\phi_E(x), \mu) \\ &= \inf_{z \in X} E_T(\phi_E(x)(z), \mu(z)) \\ &= \inf_{z \in X} E_T(E(x, z), \mu(z)) \leq \mu(x).\end{aligned}$$

Since $\mu(x) = F(y, x)$,

$$\inf_{z \in X} E_T(E(x, z), \mu(z)) = \inf_{z \in X} E_T(F(x, z), F(y, z))$$

$$\geq \inf_{z \in Y} E_T(F(x,z), F(y,z))$$
$$= F(x,y) = \mu(x)$$

since the columns of a T-indistinguishability operator are a generating family.

Therefore

$$\phi_{E_T} \circ \phi_E(x)(\mu) = \mu(x) = x^{**}(\mu).$$

$\Leftarrow$) Let us suppose that

$$\phi_{E_T} \circ \phi_E(x)(\mu) = \inf_{z \in X} E_T(E(x,z), \mu(z)) = \mu(x) \ \forall x \in X.$$

Then, for all $z \in X$ we have

$$\overrightarrow{T}(\mu(z)|E(z,x)) \geq E_T(E(x,z), \mu(z)) \geq \mu(x)$$

and therefore

$$T(\mu(x), \mu(z)) \leq E(x,z).$$

4

Isometries between Indistinguishability Operators

If we look at the definition and properties of T-indistinguishability operators, we can see that they show very metric behaviour. This is because they are a special case of a more general structure called Generalized metric spaces. Generalized metric spaces were introduced by E. Trillas ([133],[3]) as a general framework for dealing with different concepts of distance appearing in places such as metric spaces, probabilistic metric spaces, lattice metrics, etc. The idea is to valuate the map by defining the 'distance' between objects in an ordered semigroup, such that they are defined as follows:

Definition 4.1. *Let X be a set, $(M, \circ, \leq)$ an ordered semi group with identity element e and m a map $m : X \times X \to M$. (X, m) is called a generalized metric space and m a generalized metric on X if and only if for all $x, y, z \in X$*

1. $m(x, x) = e$
2. $m(x, y) = m(y, x)$
3. $m(x, y) \circ m(y, z) \geq m(x, z)$.

m separates points if and only if

$$m(x, y) = e \text{ implies } x = y.$$

Metric spaces are of course generalized metric spaces with $(M, \circ, \leq) = (\mathbb{R}^+, +, \leq)$.

Generalized metric spaces play a very important (hidden) role in fuzzy logic because the interval $[0, 1]$ with a t-norm T is an ordered semi group. 1 is the identity element and the order $\leq_T$ associated with T is the reverse of the usual order, such that 1 is the smallest element and 0 the greatest one. Thus, we can see that a set X with a T-indistinguishability operator E is nothing but a generalized metric space valued on $([0, 1], T, \leq_T)$. This encapsulates the very intuitive idea that two objects are similar, equivalent or indistinguishable when they are close and allow us to look at T-indistinguishability operators as similarities and distances at the same time.

J. Recasens: Indistinguishability Operator, STUDFUZZ 260, pp. 81–95.
springerlink.com

Another well-known object in fuzzy logic, S-metrics [139], are also generalized metric spaces. Here, the semi group is the unit interval with a t-conorm S and the usual order.

This chapter will examine some geometric aspects of T-indistinguishability operators specifically, some kinds of homomorphism between them, where a homomorphism is defined by:

Definition 4.2. *Given two t-norms T, T', a T-indistinguishability operator E on a set X and T'-indistinguishability E' on X', a morphism φ between E and E' is a pair of maps $\varphi = (h, f)$ such that the following diagram is commutative*

$$\begin{array}{ccc} X \times X & \xrightarrow{E} & [0,1] \\ \Big\downarrow{\scriptstyle h\times h} & & \Big\downarrow{\scriptstyle f} \\ X' \times X' & \xrightarrow{E'} & [0,1] \end{array}$$

(i.e. $f(E(x,y)) = E'(h(x), h(y))$ for all $x, y \in X$).
When h and f are bijective maps, φ is called an isomorphism.

Of all the possible types of homomorphism between indistinguishability operators, we will be focused on the following special cases:

a) The maps f that transform a T-indistinguishability operator E into another such operator with respect to same t-norm T. In the preceding notation, this means taking $X' = X$, $h =$ identity and $T = T'$. This case will be analyzed in Section 4.1.
b) The maps f that transform a T-indistinguishability operator E into E', another such operator with respect to a different t-norm T'. Section 4.2 looks at this more general case, taking $X = X'$, $h =$id and considering an isomorphism f between T and T'. Special attention is paid to the generators or extensional fuzzy subsets of both E and E'.

Section 4.3 studies some aspects of the group associated to a T-indistinguishability operator E on X (i.e., the group of all bijective maps $h : X \to X$ such that $E(x,y) = E(h(x), h(y)) \quad \forall x, y \in X$).

The last section relates the T-indistinguishability operators to distances in two ways.

- By relating them to S-metrics through strong negations, that are isomorphisms between T-indistinguishability operators and S-metrics.
- When T is a continuous Archimedean t-norm, the additive generators of T give isomorphisms between T-indistinguishability operators and usual distances.

4.1 Maps between T-Indistinguishability Operators

In this section the maps in the unit interval changing a T-indistinguishability operator into another one with respect to the same t-norm will be analyzed.

Definition 4.3. *A metric transform is a sub-additive and non-decreasing map* $s : [0, \infty) \to [0, \infty)$ *with* $s(0) = 0$.

Lemma 4.4. *A map* $s : [0, \infty) \to [0, \infty)$ *with* $s(0) = 0$ *is a metric transform if and only if for all* $x, y, z \in [0, 1]$ *such that* $x + y \geq z$, $s(x) + s(y) \geq s(z)$.

Proof. $\Rightarrow$) If s is a metric transform, then for $x + y \geq z$

$$s(x) + s(y) \geq s(x + y) \geq s(z).$$

$\Leftarrow$) Taking $x + y = z$, we have $s(x) + s(y) \geq s(z)$. Since $s(0) = 0$, s is sub-additive.

On the other hand, if $x \geq y$, then $x + 0 \geq y$ and $s(x) + s(0) \geq s(y)$. Therefore, s is a non-decreasing map.

Proposition 4.5. *Let E be a T-indistinguishability operator on a set X with T a continuous Archimedean t-norm with an additive generator t and f a map $f : [0, 1] \to [0, 1]$. $f \circ E$ is a T-indistinguishability operator on X if and only if there exists a metric transform s such that the restriction $f_{|\mathrm{Im}(E)}$ of f to the image of E satisfies*

$$f_{|\mathrm{Im}(E)} = t^{[-1]} \circ s \circ t.$$

Proof. Given $x, y, z \in X$, let $E(x, y) = a$, $E(y, z) = b$, $E(x, z) = c$, $t(a) = u$, $t(b) = v$, $t(c) = w$.

$\Rightarrow$) If $f \circ E$ is a T-indistinguishability operator, then $(f \circ E)(x, x) = 1$ and therefore $f(1) = 1$.

1) If T is strict Archimedean, then $t(0) = +\infty$, $t^{[-1]}(x) = t^{-1}(x)$ and, since E is T-transitive, $t^{-1}(t(a) + t(b)) \leq c$, or $t(a) + t(b) \geq t(c)$.
   In a similar way, $t \circ f(a) + t \circ f(b) \geq t \circ f(c)$.
   Therefore, if $u + v \geq w$, then

$$t \circ f \circ t^{-1}(u) + t \circ f \circ t^{-1}(v) \geq t \circ f \circ t^{-1}(w).$$

   Defining $s = t \circ f \circ t^{-1}$, it follows that $s(0) = 0$ and if $u, v, w \in \mathrm{Im}(E)$ and $u + v \geq w$, then $s(u) + s(v) \geq s(w)$.
2) If T is non-strict and t is an additive generator of T, then $t^{[-1]}(x) = 0$ if $x > t(0)$ and $t^{[-1]}(x) = t^{-1}(x)$ if $x \leq t(0)$.
   If $t(a) + t(b) \leq t(0)$, then $t(a) + t(b) \geq t(c)$.
   If $t(a) + t(b) > t(0)$, since $t(c) \leq t(0)$ we also have $t(a) + t(b) \geq t(c)$.
   Therefore, in any case $t(a) + t(b) \geq t(c)$.
   In a similar way, $t \circ f(a) + t \circ f(b) \geq t \circ f(c)$.

If $u, v, w \in \mathrm{Im}(E)$ and $u + v \geq w$, then

$$t \circ f \circ t^{[-1]}(u) + t \circ f \circ t^{[-1]}(v) \geq t \circ f \circ t^{[-1]}(w).$$

Taking $t \circ f \circ t^{[-1]} = s$, then $f = t^{[-1]} \circ s \circ t$.

$\Leftarrow$) If $f = t^{[-1]} \circ s \circ t$ with s a metric transform, then f is a non-decreasing map with $f(1) = 1$. Let us show that $f \circ E$ is a T-indistinguishability operator:

- Reflexivity. $f \circ E(x, x) = f(1) = 1$.
- Symmetry is trivial.
- Transitivity.
  a) If T is a strict continuous Archimedean t-norm, then $t(a) + t(b) \geq t(c)$.

$$\begin{aligned} T(f(a), f(b)) &= t^{-1}(t \circ f(a) + t \circ f(b)) \\ &= t^{-1}\left(t \circ t^{-1} \circ s \circ t(a) + t \circ t^{-1} \circ s \circ t(b)\right) \\ &= t^{-1}(s \circ t(a) + s \circ t(b)) \leq t^{-1} \circ s \circ t(c) = f(c) \end{aligned}$$

  and therefore $f \circ E$ is T-transitive.
  b) If T is a non-strict continuous Archimedean t-norm, then

$$T(f(a), f(b)) = t^{[-1]}(t(t^{[-1]} \circ s \circ t)(a)) + t(t^{[-1]} \circ s \circ t)(b))).$$

  If $(s \circ t)(a) > t(0)$ or $(s \circ t)(b) > t(0)$, then $\left(t^{[-1]} \circ s \circ t\right)(a) = 0$ or $\left(t^{[-1]} \circ s \circ t\right)(b) = 0$ and therefore $T(f(a), f(b)) \leq f(c)$.
  Otherwise, we also get the inequality with the same arguments as in a).

Example 4.6. Let $s : [0, \infty) \to [0, \infty)$ be the map defined by $s(x) = x^\alpha$ for all $x \in [0, \infty)$ with $0 < \alpha \leq 1$. s is a metric transform, since $s(0) = 0$ and

$$s(x + y) = (x + y)^\alpha \leq x^\alpha + y^\alpha = s(x) + s(y).$$

If E is a T-indistinguishability operator on a set X with T the Łukasiewicz t-norm, $t(x) = 1 - x$ an additive generator, then

$$E'(x, y) = f(E(x, y)) = (t^{-1} \circ s \circ t)(E(x, y)) = 1 - (1 - E(x, y))^\alpha$$

is also a T-indistinguishability operator on X.

Let T be the Product t-norm and $t(x) = -\ln x$ an additive generator of T. If E is a T-indistinguishability operator on a set X, then $E'(x, y) = e^{-(-\ln E(x,y))^\alpha}$ also is a T-indistinguishability operator on X.

Let us now study the maps that preserve min-indistinguishability operators.

Lemma 4.7. *Let E be a* min-*indistinguishability operator on a set X. If for $x, y, z \in X$ $E(x, y) \leq E(y, z) \leq E(x, z)$, then $E(x, y) = E(y, z)$.*

Proof. Taking $E(x,y) = a$, $E(y,z) = b$, $E(x,z) = c$, due to the min-transitivity we have $\min(a,b) \leq c$, $\min(b,c) \leq a$ and $\min(a,c) \leq b$.

If $a \leq b \leq c$, the last two inequalities implies $b \leq a$ and $a \leq b$ and therefore $a = b$.

Proposition 4.8. *Let* E *be a* min*-indistinguishability operator on a set* X *and* $f : [0,1] \to [0,1]$ *a map in the unit interval.* $f \circ E$ *is a* min*-indistinguishability operator on* X *if and only if* $f(1) = 1$ *and* f *restricted to* $\mathrm{Im}(E)$ *is a non-decreasing function.*

Proof. If E is a min-indistinguishability operator, then given $x, y, z \in X$, we can assume without loss of generality, and thanks to the preceding lemma, that

$$E(x,y) = E(y,z) = a \leq E(x,z) = c.$$

$\Rightarrow$) Since $f \circ E$ is also min-transitive, then $f(a) \leq f(c)$ and f is non-decreasing on Im E. On the other hand, $f(1) = f(E(x,x)) = 1$.

$\Leftarrow$) $f(1) = 1$ implies the reflexivity on $f \circ E$ and symmetry is trivially satisfied.

Since

$$a = \min(E(x,y), E(y,z)) \leq E(x,z) = c$$

and f is a non-decreasing map on Im E,

$$f(a) = \min(f \circ E(x,y), f \circ E(y,z)) \leq f \circ E(x,z).$$

Since

$$a = \min(E(x,z), E(x,y)) = E(y,z) = a$$

and also

$$a = \min(E(x,z), E(y,z)) = E(x,y) = a,$$

in both cases min-transitivity is also satisfied by $f \circ E$.

4.2 Indistinguishability Operators and Isomorphic t-Norms

In this section, the relation between indistinguishability operators with respect to isomorphic t-norms will be studied.

Definition 4.9. *Two continuous t-norms* T, T' *are isomorphic if and only if there exists a bijective map* $f : [0,1] \to [0,1]$ *such that* $f \circ T = T' \circ (f \times f)$.

Isomorphisms f are continuous and increasing maps.

It is well known that all strict continuous Archimedean t-norms are isomorphic. In particular, they are isomorphic to the Product t-norm.

Also, all non-strict continuous Archimedean t-norms are isomorphic. In particular, they are isomorphic to the Łukasiewicz t-norm.

The next proposition relates the isomorphisms of continuous Archimedean t-norms with their additive generators.

Proposition 4.10. *Let f be a bijective map $f : [0,1] \to [0,1]$, T, T' two continuous Archimedean t-norms and t, t' additive generators of T and T' respectively. If f is an isomorphism between T and T', then there exists $\alpha \in (0,1]$ such that $f = t'^{[-1]}(\alpha t)$.*

Proof. $\forall x, y \in [0,1]$, $f(T(x,y)) = T'(f(x), f(y))$
or equivalently

$$\begin{aligned} f \circ t^{[-1]}(t(x) + t(y)) &= t'^{[-1]}((t' \circ f)(x) + (t' \circ f)(y)) \\ t^{[-1]}(t(x) + t(y)) &= (f^{-1} \circ t'^{[-1]})((t' \circ f)(x) + (t' \circ f)(y) \end{aligned}$$

which means that $t' \circ f$ is also a generator of T. Since two generators of a t-norm differ only by a multiplicative positive constant, $t = k(t' \circ f)$ with $k > 0$ and putting $\alpha = 1/k$,

$$f = t'^{[-1]} \alpha t.$$

Example 4.11. The only automorphism of the Łukasiewicz t-norm is the identity map.

Indeed, taking $t(x) = 1 - x$, then $f(x) = 1 - \alpha + \alpha\, x$ and the only bijective linear map in [0,1] is the identity.

The automorphisms of the Product t-norm are $f(x) = x^\alpha$ with $\alpha > 0$.

More general, the only automorphism of a non-strict Archimedean t-norm is the identity map and for strict t-norms, every $\alpha > 0$ produces an isomorphism f_α with $f_\alpha \neq f_\beta$ if $\alpha \neq \beta$.

Lemma 4.12. *If T, T' are two isomorphic t-norms, then their residuations $\overrightarrow{T}, \overrightarrow{T'}$ also are isomorphic.*

Proof. If $f \circ T = T' \circ (f \times f)$, then for all $x, y \in [0,1]$,

$$\begin{aligned} f \circ \overrightarrow{T}(x|y) &= f(\sup\{\alpha \in [0,1] \,|T(\alpha, x) \leq y\}) \\ &= \sup\{f(\alpha) \in [0,1] \,|f^{-1} \circ T'(f(\alpha), f(x)) \leq y\} \\ &= \sup\{f(\alpha) \in [0,1] \,|T'(f(\alpha), f(x)) \leq f(y)\} \\ &= \overrightarrow{T'}(f(x)|f(y)). \end{aligned}$$

Proposition 4.13. *If E is a T-indistinguishability operator on a set X for a given t-norm T and f is a continuous, increasing and bijective map $f : [0,1] \to [0,1]$, then $f \circ E$ is a T'-indistinguishability operator with $T' = f \circ T \circ (f^{-1} \times f^{-1})$.*

Proof. $f \circ E$ is trivially a reflexive and symmetric fuzzy relation on X. Since E is T-transitive, for all $x, y, z \in X$ it holds

$$T(E(x,y), E(y,z)) \leq E(x,z).$$

From this,

$$T(f^{-1} \circ f \circ E(x,y), f^{-1} \circ f \circ E(y,z)) \leq f^{-1} \circ f \circ E(x,z),$$

$$f \circ T(f^{-1} \times f^{-1})(f \circ E(x,y), f \circ E(y,z)) \leq f \circ E(x,z)$$

and $f \circ E$ is T'-transitive.

Under the assumptions of the preceding proposition, E and $f \circ E$ will be called similar indistinguishability operators.

Example 4.14. Let E be a T-indistinguishability operator on a set X with respect to a t-norm T and $f_\alpha(x) = x^\alpha$ for some $\alpha > 0$. Then $f_\alpha \circ E$ is a T_α-indistinguishability operator on X with

$$T_\alpha(x,y) = \left(T\left(x^{\frac{1}{\alpha}}, y^{\frac{1}{\alpha}}\right)\right)^\alpha.$$

In particular, if T is the Product t-norm, then T' is also the Product t-norm and if $T = \min$, then $T' = \min$ (cf. Section 4.1).

If T is the Łukasiewicz t-norm, then $(T_\alpha)_{\alpha>0}$ is the Schweizer-Sklar family of t-norms

$$T_\alpha(x,y) = \left(\max\left(x^{\frac{1}{\alpha}} + y^{\frac{1}{\alpha}} - 1, 0\right)\right)^\alpha$$

Therefore, given a T_α-indistinguishability operator E_α, it is easy to find a similar T_β-indistinguishability operator E_β, $(\alpha, \beta > 0)$.

Example 4.15. Let E be a T-indistinguishability operator on a set X with respect to a non-strict continuous Archimedean t-norm T with normalized additive generator t. Taking $f(x) = 1 - (t(x))^\alpha$, $\alpha > 0$, then

$$f \circ E(x,y) = 1 - (t(E(x,y)))^\alpha$$

is a T_α-indistinguishability operator where T_α is the non-strict Archimedean t-norm

$$T_\alpha(x,y) = 1 - \min((1 - x^{\frac{1}{\alpha}}) + (1-y)^{\frac{1}{\alpha}}, 1)^\alpha.$$

Let us observe that, in this case, the family $\{T_\alpha\}_{\alpha>0}$, known as Yager family, is independent from the t-norm T, more precisely, from the generator t.

In particular, for $\alpha = 1$, T_α is the Łukasiewicz t-norm, and the operator $E_1 = 1 - t(E)$ is always T_1-transitive.

Lemma 4.16. *Let T be a continuous t-norm and E a T-indistinguishability operator on a set X. If μ is a generator of E and f a continuous, increasing and bijective map $f : [0,1] \to [0,1]$, then $f \circ \mu$ is a generator of the similar T'-indistinguishability operator $f \circ E$.*

Proof. $\forall x, y \in X, E(x,y) \leq E_\mu(x,y) = \overrightarrow{T}(\max(\mu(x),\mu(y))|\min(\mu(x),\mu(y)))$.

$$\begin{aligned} f \circ E(x,y) &\leq f \circ \overrightarrow{T}(\max(\mu(x),\mu(y))|\min(\mu(x),\mu(y))) \\ &= f \circ \overrightarrow{T}(f^{-1} \times f^{-1})(f \circ \max(\mu(x),\mu(y))|f \circ \min(\mu(x),\mu(y))) \\ &= \overrightarrow{T'}(\max(f \circ \mu(x), f \circ \mu(y))|\min(f \circ \mu(x), f \circ \mu(y))) \end{aligned}$$

which means that $f \circ \mu$ is a generator of $f \circ E$.

In a similar way, the next result can be proved.

Proposition 4.17. *Let T be a continuous t-norm and E a T-indistinguishability operator on a set X. If $(\mu_i)_{i\in I}$ is a generating family of E and f a continuous, increasing and bijective map $f : [0,1] \to [0,1]$, then $(f \circ \mu_i)_{i\in I}$ is a generating family of the similar T'-indistinguishability operator $f \circ E$.*

Since the relation of being similar is symmetric, the following two results hold:

Corollary 4.18. *Similar indistinguishability operators have the same dimension.*

Corollary 4.19. *With the preceding notations, $(\mu_i)_{i\in I}$ is a basis of E if and only if $(f \circ \mu_i)_{i\in I}$ is a basis of $f \circ E$.*

4.3 Isometries between Indistinguishability Operators

Definition 4.20. *Given two sets X, Y and two T-indistinguishability operators E, F on X, Y respectively, a map $\tau : X \to Y$ is an isometry if and only if $E(x,y) = F(\tau(x), \tau(y))$ $\forall x, y \in X$.*

Lemma 4.21. *With the previous notations, given an isometry $\tau : X \to Y$, if F separates points, then E also separates points and τ is injective.*

Proof. Trivial.

Lemma 4.22. *Given a T-indistinguishability operator E on X, let us consider the crisp relation on $X : x \sim y$ if and only if $E(x,y) = 1$. $\sim$ is an equivalence relation and if μ is a generator of E, then the fuzzy relation $\overline{E}$ defined by $\overline{E}(\bar{x}, \bar{y}) = E(x,y)$ is a T-indistinguishability operator in the quotient set $\overline{X} = X/\sim$ that separates points and $\bar{\mu}$ defined by $\bar{\mu}(\bar{x}) = \mu(x)$ is a generator of $\overline{E}$.*

Proof. Straightforward.

Before studying the isometries between indistinguishability operators, it is convenient to analyze when two fuzzy sets μ, ν generate the same operator (i.e. $E_\mu = E_\nu$).

Theorem 4.23. *Let T be a continuous Archimedean t-norm, t a generator of T and μ, ν fuzzy subsets of X. $E_\mu = E_\nu$ if and only if $\forall x \in X$ one of the following conditions holds:*

a) $t(\mu(x)) = t(\nu(x)) + k_1$ *with* $k_1 \geq \sup\{-t(\nu(x)) | x \in X\}$
or
b) $t(\mu(x)) = -t(\nu(x)) + k_2$ *with* $k_2 \geq \sup\{t(\nu(x)) | x \in X\}$.

Moreover, if T is non-strict, then $k_1 \leq \inf\{t(0) - t(\nu(x)) \mid x \in X\}$ *and* $k_2 \leq \inf\{t(0) + t(\nu(x)) \mid x \in X\}$.

Proof. $\Rightarrow$) Due to Lemma 4.22 we can suppose that μ is a one to one map.

$$\begin{aligned}E_\mu(x,y) &= \overrightarrow{T}(\max(\mu(x),\mu(y)) | \min(\mu(x),\mu(y))) \\ &= t^{-1}(t(\min(\mu(x),\mu(y))) - t(\max(\mu(x),\mu(y)))). \\ E_\nu(x,y) &= t^{-1}(t(\min(\nu(x),\nu(y))) - t(\max(\nu(x),\nu(y))))\end{aligned}$$

where $t^{[-1]}$ is replaced by t^{-1} because all the values in brackets are between 0 and $t(0)$.

If $E_\mu = E_\nu$, then

$$\begin{aligned}&t(\min(\mu(x),\mu(y))) - t(\max(\mu(x),\mu(y))) \\ &= t(\min(\nu(x),\nu(y))) - (t\max(\nu(x),\nu(y))).\end{aligned}$$

Therefore, $t(\mu(x)) - t(\mu(y)) = t(\nu(x)) - t(\nu(y))$ or $t(\mu(y)) - t(\mu(x)) = t(\nu(x)) - t(\nu(y))$.

Let us fix $y_0 \in X$ and consider the map $M(x) = t(\mu(x)) - t(\mu(y_0))$.

Taking $k' = t(\nu(y_0))$, then $M(x) = t(\nu(x)) - k'$ or $M(x) = -t(\nu(x)) + k'$.

We need to prove that there does not exist $x, y \in X$ with

$$M(x) = t(\nu(x)) - k \text{ and } M(y) = -t(\nu(y)) + k \qquad (*)$$

Suppose that there exist $x, y \in X$ such that both equalities (*) hold. Since $M(y_0) = 0$, we can take $x \neq y_0$ and $y \neq y_0$, in this case

$$\begin{aligned}E_\mu(x,y) &= t^{-1}(t(\min(\mu(x),\mu(y))) - t(\max(\mu(x),\mu(y)))) \\ &= t^{-1}(t(\min(\mu(x),\mu(y))) - t(\mu(y_0)) - (t(\max(\mu(x),\mu(y))) - t(\mu(y_0)))) \\ &= t^{-1}(\max(M(x),M(y)) - \min(M(x),M(y))).\end{aligned}$$

$E_\nu(x,y) = t^{-1}(t(\min(\nu(x),\nu(y))) - t(\max(\nu(x),\nu(y))))$ and since $M(x) + M(y) = t(\nu(x)) - t(\nu(y))$, either

$$M(x) - M(y) = M(x) + M(y) \text{ and } y = y_0$$

or

$$M(x) - M(y) = -(M(x) + M(y)) \text{ and } x = y_0.$$

So

$$t(\mu(x)) = t(\nu(x)) - t(\nu(y_0)) + t(\mu(y_0)) = t(\nu(x)) + k_1$$

or

$$t(\mu(x)) = -t(\nu(x)) + t(\nu(y_0)) + t(\mu(y_0)) = -t(\nu(x)) + k_2.$$

$\Leftarrow$) Trivial.

Example 4.24. If T is the Łukasiewicz t-norm, with the previous notations

$$\mu(x) = \nu(x) + k \text{ with } \inf\{1 - \nu(x)\} \geq k \geq \sup_{x \in X}\{-\nu(x)\}$$

or

$$\mu(x) = -\nu(x) + k \text{ with } \inf_{x \in X}\{1 + \nu(x)\} \geq k \geq \sup_{x \in X}\{\nu(x)\}.$$

Indeed, taking $t(x) = 1 - x$,

a)

$$1 - \mu(x) = 1 - \nu(x) + k_1$$

with

$$\sup\{-1 + \nu(x) | x \in X\} \leq k_1 \leq \inf\{\nu(x) | x \in X\}$$

and therefore

$$\mu(x) = \nu(x) + k$$

with

$$\inf_{x \in X}\{1 - \nu(x) \geq k \geq \sup_{x \in X}\{-\nu(x)\}$$

or

b)

$$1 - \mu(x) = -1 + \nu(x) + k_2$$

with

$$\sup\{1 - \nu(x) \mid x \in X \} \leq k_2 \leq \inf\{2 - \nu(x) | x \in X\}$$

and therefore

$$\mu(x) = -\nu(x) + k$$

with

$$\inf_{x \in X}\{1 + \nu(x)\} \geq k \geq \sup_{x \in X}\{\nu(x)\}.$$

Example 4.25. If T is the product t-norm, then

$$\mu(x) = \frac{\nu(x)}{k} \text{ with } k \geq \sup_{x \in X}\{\nu(x)\}$$

or

$$\mu(x) = \frac{k}{\nu(x)} \text{ with } k \leq \inf_{x \in X}\{\nu(x)\}.$$

Indeed, taking $t(x) = -\ln x$,

a)

$$-\ln \mu(x) = -\ln \nu(x) + k_1$$

with

$$k_1 \geq \sup\{\ln \nu(x) \mid x \in X\}$$

and therefore

$$\mu(x) = \frac{\nu(x)}{k}$$

with

$$k \geq \sup\{\nu(x)\}$$

or

b)

$$-\ln \nu(x) = \ln \nu(x) + k_2$$

with

$$k_2 \geq \sup\{-\ln \nu(x) | x \in X\}$$

and therefore

$$\mu(x) = \frac{k}{\nu(x)}$$

with

$$k \leq \inf_{x \in X}\{\nu(x)\}.$$

Theorem 4.26. *Let T be the t-norm minimum and let μ be a fuzzy subset of X such that there exists an element x_M of X with $\mu(x_M) \geq \mu(x)$ $\forall x \in X$. Let $Y \subset X$ be the set of elements x of X with $\mu(x) = \mu(x_M)$ and $s = \sup\{\mu(x)$ such that $x \in X - Y\}$. A fuzzy subset ν of X generates the same T-indistinguishability operator than μ if and only if*

$$\forall x \in X - Y \;\; \mu(x) = \nu(x) \text{ and } \nu(y) = t \text{ with } s \leq t \leq 1 \;\forall y \in Y.$$

Proof. It follows easily from the fact that

$$E_\mu(x,y) = \begin{cases} \min(\mu(x), \mu(y)) & \text{if } \mu(x) \neq \mu(y) \\ 1 & \text{if } \mu(x) = \mu(y). \end{cases}$$

The next theorem relates the previous results with the group of isometries of a set X equipped with an indistinguishability operator.

Theorem 4.27. *Let T be a continuous t-norm and E_μ the T-indistinguishability operator on X generated by the fuzzy subset μ of X. The map $\tau : X \to X$ is an isometry if and only if there exists a fuzzy subset ν of X with $E_\mu = E_\nu$ and $\mu \circ \tau = \nu$.*

Proof

$$\begin{aligned} E_\mu(x,y) &= E_\mu(\tau(x),\tau(y)) = \\ &= E_T(\mu\circ\tau(x),\mu\circ\tau(y)) = E_{\mu\circ\tau}(x,y). \end{aligned}$$

Corollary 4.28. *Let T be a continuous t-norm and E_μ, E_ν two T-indistinguishability operators on X, Y respectively generated by μ and ν. A bijective map $\tau : X \to Y$ is an isometry if and only if $\mu = \upsilon \circ \tau$.*

4.3.1 The Group of Isometries of ([0,1], E_T)

The following universal property of the natural indistinguishability operator E_T gives special interest to its study.

Lemma 4.29. *Let T be a continuous t-norm, E_μ the T-indistinguishability operator on X generated by the one to one fuzzy subset μ of X and E_T the natural T-indistinguishability operator on $[0,1]$. The membership function μ is an isometry of X into $[0,1]$.*

Proof. Trivial.

In fact, the preceding results can be translated to the language of categories in such a way that E_T is a final universal object:

Fixing a continuous t-norm T, let us consider the category C whose objects are the pairs (X,μ) consisting of a set X and a fuzzy subset μ of X and with morphisms $\tau : (X,\mu) \to (Y,\nu)$, the isometries of $(X,E_\mu) \to (Y,E_\nu)$ with $\upsilon \circ \tau = \mu$ (so that the set of morphisms between two objects is either empty or it contains only one map).

It is clear that (X,μ) and (Y,ν) are isomorphic if and only if $E_\mu = E_\nu$, and Lemma 4.29 can now be expressed in the following way.

Proposition 4.30. *In the category C, $([0,1],\mathrm{id})$ is a final universal object.*

Let us calculate the group of isometries of $([0,1],E_T)$.

Theorem 4.31. *Let T be a non-strict continuous Archimedean t-norm T and t an additive generator of T. The group of isometries of $([0,1],E_T)$ consists of the identity and the strong negation generated by t.*

Proof. E_T is generated by the identity map id. From Theorem 4.23, using the same notations and taking ν=id and X=[0,1], if follows that $k_1 = 0$ and $k_2 = t(0)$, so that in the first case μ=id and in the second case $\mu(x) = t^{-1}(t(0) - t(x))$. Since it has been shown that these are the only fuzzy sets of [0,1] that generate E_T, the result follows immediately from Theorem 4.27.

Definition 4.32. *Given a t-norm T and $a \in [0,1]$, the map $t_a : [0,1] \to [0,1]$ defined by $t_a(x) = T(a,x)$ will be called the T-translation by a.*

Theorem 4.33. *Let T be a strict continuous Archimedean t-norm. The group of isometries of $([0,1], E_T)$ is the set of T-translations of $[0,1]$ (i.e. $\{t_a | a \in [0,1]\}$).*

Proof. Taking ν=id and $X = [0,1]$ in Theorem 4.23, it follows that $k_1 \geq 0$ and the other case cannot occur, since it would imply $k_2 = \infty$.

So $t(\mu(x)) = t(x) + t(a)$, where $t(a) = k_1$ and $\mu(x) = T(x,a)$.

Since $t^{-1} : [0,\infty] \to [0,1]$ is onto, every translation generates E_T.

Once shown that the set of generators of E_T consists of the set of translations, the result follows from Theorem 4.27.

Theorem 4.34. *Let T be the t-norm minimum. The group of isometries of $([0,1], E_T)$ consists of only the identity map.*

Proof. It follows immediately from Theorem 4.26 and Theorem 4.27.

As a corollary of these theorems, we can answer an interesting question relating indistinguishability operators and negations. In fuzzy logic E_T can be viewed as the biimplication in the sense that $E_T(x,y)$ expresses the degree of equivalence between x and y. It seem reasonable to demand that the degree of equivalence between elements coincide with the degree of their negations (i.e. $E_T(x,y) = E_T(\varphi(x), \varphi(y))$). The following question arises therefore naturally: When a negation is an isometry of E_T? The answer is given in the following corollary.

Corollary 4.35. *Given a continuous Archimedean t-norm T and a strong negation φ, φ is an isometry of E_T if and only if T is non-strict and T and φ have a common generator.*

4.4 T-Indistinguishability Operators and Distances

There are basically two ways of relating T-indistinguishability operators and distances.

1. If φ is a strong negation, S the dual t-conorm of T with respect to φ and E a T-indistinguishability operator on a set X, then $m = \varphi \circ E$ is an S-metric on X (see Definition 4.37). In particular, if T is greater than or equal to the Łukasiewicz t-norm, then m is a pseudodistance on X that is a distance if and only if E separates points. If T is the minimum t-norm, then m is a pseudoultrametric as will be seen in Chapter 5.
2. If T is a continuous Archimedean t-norm and t an additive generator of T, then E is a T-indistinguishability operator on a set X if and only if $d = t \circ E$ is a pseudodistance on X and E separates points if and only if d is a distance on X.

4.4.1 T-Indistinguishability Operators and S-Metrics

Definition 4.36. *Let T be a t-norm and φ a strong negation. Then $S = \varphi \circ T \circ \varphi$ is the dual t-conorm of T with respect to S. In this case (T, S, φ) is called a De Morgan triplet.*

Definition 4.37. *Given a t-conorm S and a set X a map $m : X \times X \to [0, 1]$ is an S-pseudometric on X if and only if for all $x, y, z \in X$*

1. $m(x, x) = 0$
2. $m(x, y) = m(y, x)$
3. $S(m(x, y), m(y, z)) \geq m(x, z)$.

m is an S-metric if and only if it also satisfies

$$m(x, y) = 0 \text{ implies } x = y.$$

S-pseudometrics and T-indistinguishability operators are dual concepts. For instance, there is a Representation Theorem for S-pseudometrics dual to that of T-indistinguishability operators and generating families, dimension and basis can be also defined. Also from an irreflexive and symmetric fuzzy relation R the greatest S-metric among the ones smaller than or equal to R can be built using the $\inf -S$ product.

Proposition 4.38. *Let S, S' be two t-conorms with $S \leq S'$. If m is an S-pseudometric on a set X, then m is also an S'-pseudometric.*

Proof. Trivial.

Proposition 4.39. *Let (T, S, φ) be a De Morgan triplet and X a set. E is a T-indistinguishability operator on X if and only if $m = \varphi \circ E$ is an S-pseudometric on X. E separates points if and only if m is an S-metric.*

Proof. Trivial.

In fact, this last proposition states that φ is an isomorphism (of Generalized metric spaces) between E and m.

Corollary 4.40. *Let T be a t-norm greater than or equal to the t-norm of Łukasiewicz. E is a T-indistinguishability on a set X if and only if $m = \varphi \circ E$ is a pseudodistance on X. E separates points if and only if m is a distance on X.*

4.4.2 T-Indistinguishability Operators and Distances

Proposition 4.41. *Let T be a continuous Archimedean t-norm and t an additive generator of T.*

1. *If d is a pseudo distance on a set X, then $E = t^{[-1]} \circ d$ is a T-indistinguishability operator on X.*
2. *If E is a T-indistinguishability on X, then $d = t \circ E$ is a pseudo distance on X.*

d is a distance on X if and only if E separates points.

Proof

- Reflexivity. $d(x, y) = 0$ if and only if $t^{[-1]} \circ d(x, y) = 1$.
- Symmetry. $d(x, y) = d(y, x)$ if and only if

$$E(x, y) = t^{[-1]} \circ d(x, y) = t^{[-1]} \circ d(y, x) = E(y, x).$$

- Transitivity.

$$\begin{aligned} d(x, y) + d(y, z) \geq d(x, z) &\Leftrightarrow t(E(x, y)) + t(E(y, z)) \geq t(E(x, z) \\ &\Leftrightarrow t^{[-1]} \left(t(E(x, y)) + t(E(y, z))\right) \leq E(x, z) \\ &\Leftrightarrow T(E(x, y), E(y, z)) \leq E(x, z). \end{aligned}$$

- $d(x, y) \neq 0$ when $x \neq y$ if and only if $E(x, y) = t^{[-1]} \circ d(x, y) \neq 1$ when $x \neq y$.

This bijection is not canonical but depends o the generator t. The next proposition relates the distances and indistinguishability operators generated by different additive generators of a t-norm.

Proposition 4.42. *Let T be a continuous Archimedean t-norm and t and u two additive generators of T such that $u = \alpha \cdot t$ with $\alpha > 0$.*

1. *If d is a pseudo distance on a set X, $E = t^{[-1]} \circ d$ and $E' = u^{[-1]} \circ d$, then for all $x, y \in X$, $E'(x, y) = t^{[-1]} \left(\frac{d(x,y)}{\alpha} \right)$.*
2. *If E is a T-indistinguishability operator on X, $d = t \circ E$ and $d' = u \circ E$, then for all $x, y \in X$, $d'(x, y) = \alpha \cdot t(E(x, y))$.*

Proof

1. If $u = \alpha \cdot t$, then $u^{[-1]}(x) = t^{[-1]}(\frac{x}{\alpha})$ for all $x \in [0, u(0)]$ and *1.* follows.
2. is trivial.

5

Min-indistinguishability Operators and Hierarchical Trees

min-indistinguishability operators are widely used in Taxonomy because they are closely related to hierarchical trees. Indeed, given a min-indistinguishability operator on a set X and $\alpha \in [0,1]$, the α-cuts of E are partitions of X and if $\alpha \geq \beta$, then the α-cut is a refinement of the partition corresponding to the β-cut. Therefore, E generates an indexed hierarchical tree. Reciprocally, from an indexed hierarchical tree a min-indistinguishability operator can be generated. These results follow from the fact that $1 - E$ is a pseudo ultrametric. Pseudo ultrametrics are pseudodistances where, in the triangular inequality, the addition is replaced by the more restrictive max operation. The topologies generated by ultrametrics are very peculiar, since if two balls are non disjoint, then one of them is included in the other one.

An important issue with respect to fuzzy relations is their storage, since in some applications they can be defined on sets of very large cardinality. The work [131] provides a very easy way to represent min-indistinguishability operators, which will be discussed in Section 5.2.

One important combinatorial problem is determining how many essentially different min-indistinguishability operators and therefore how many essentially different hierarchical trees there are on a finite set of cardinality n. This problem seems very difficult, and here we present the answer for $n \leq 5$.

5.1 Min-indistinguishability Operators and Ultrametrics

Definition 5.1. *A map $m : X \times X \to \mathbb{R}$ is a pseudo ultrametric if and only if for all $x, y, z \in X$*

1. $m(x,x) = 0$.
2. $m(x,y) = m(y,x)$.
3. $\max(m(x,y), m(y,z)) \geq m(x,z)$.

If $m(x,y) = 0$ implies $x = y$, then it is called an ultrametric.

J. Recasens: Indistinguishability Operator, STUDFUZZ 260, pp. 97–105.
springerlink.com

Pseudo ultrametrics are pseudo distances where the triangular inequality has been strengthen replacing the addition by the maximum.

Ultrametric spaces have a very special behaviour.

Proposition 5.2. *Let m be an ultrametric on X. Then*

1. *If $B(x, r)$ denotes the ball of centre x and radius r and $y \in B(x, r)$, then $B(x, r) = B(y, r)$. (All elements of a ball are its centre).*
2. *If two balls have non-empty intersection, then one of them is contained in the other one.*

Proof

1. If $y \in B(x, r)$, then $m(x, y) \leq r$. Now let z be in $B(x, r)$. Then $m(x, z) \leq r$ and
$$m(y, z) \leq \max(m(x, z), m(x, y)) \leq r$$
which means that $z \in B(y, r)$. If $y \in B(x, r)$, then $m(x, y) \leq r$. Now let z be in $B(y, r)$. Then $m(y, z) \leq r$, and
$$m(x, z) \leq \max(m(y, z), m(x, y)) \leq r.$$
So $z \in B(x, r)$.
2. Let $B(x, r) \cap B(y, s) \neq \emptyset$ and $s \leq r$. There exists $z \in B(x, r) \cap B(y, s)$ which means $m(z, x) \leq r$ and $m(y, z) \leq s$. Let $t \in B(y, s)$. Then $m(t, y) \leq s$.
$$m(t, x) \leq \max(m(x, z), m(z, t)) \leq \max(m(x, z), m(y, t), m(y, z)) \leq r.$$
So $t \in B(x, r)$.

Proposition 5.3. *Let E be a fuzzy relation on a set X. E is a* min*-indistinguishability operator on X if and only if $m = 1 - E$ is a pseudo ultrametric.*

Proof

- Reflexivity and symmetry are trivial.
-
$$\begin{aligned} &\max(m(x, y), m(x, z)) \geq m(x, z) \\ &\Leftrightarrow \max(1 - E(x, y), 1 - E(x, z)) \geq 1 - E(x, z) \\ &\Leftrightarrow 1 - \min(E(x, y), E(x, z)) \geq 1 - E(x, z) \\ &\Leftrightarrow \min(E(x, y), E(x, z)) \leq E(x, z). \end{aligned}$$

Grom this proposition interesting results follow. One of them is that the cardinality of $\text{Im}(E) = \{E(x, y)\}$ is smaller than or equal to the cardinality of X. In particular, if X is finite of cardinality n and E is identified with a matrix, then the number of different entries of the matrix is less or equal than n, which simplifies calculation and storage.

Another important result relates to the α-cuts. Let us recall that for a fuzzy relation on X and $\alpha \in [0,1]$, the α-cut of E is the set E_α of pairs $(x,y) \in X \times X$ such that $E(x,y) \geq \alpha$. Of course, if $\alpha \geq \beta$, then $E_\alpha \leq E_\beta$.

Proposition 5.4. *Let E be a fuzzy relation on X. E is a* min-*indistinguishability operator on X if and only if for each $\alpha \in [0,1]$, the α-cut of E is an equivalence relation on X.*

Proof. $\Rightarrow$) Trivial.

$\Leftarrow$) If $E(x,y) = \alpha$ and $E(y,z) = \beta$ with $\alpha \leq \beta$, (x,y) and (y,z) belong to the α-cut of E, which is an equivalence relation. Therefore (x,z) also belongs to the α-cut of E and $\min(E(x,y), E(y,z)) \leq E(x,z)$.

In fact, the last proposition also follows from Propositions 5.2 and 5.3.

5.2 Min-indistinguishability Operators and Hierarchical Trees

Definition 5.5. *A hierarchical tree of a finite set X is a sequence of partitions $A_1, A_2, ..., A_k$ of X such that every partition refines the preceding one.*

A hierarchical tree is indexed if every partition A_i has associated a non-negative number λ_i and $\lambda_i < \lambda_{i+1}$ for all $i = 1, 2, ..., k-1$.

Proposition 5.6. *Every* min-*indistinguishability operator E on a finite set X generates an indexed hierarchical tree on X.*

Proof. We can consider the α-cuts of E with $\alpha \in \mathrm{Im}(E)$.

Reciprocally,

Proposition 5.7. *Every indexed hierarchical tree $A_1, A_2, ..., A_k$ of a finite set X with $\lambda_k = 1$ generates a* min-*indistinguishability operator E on X.*

Proof. If $A_1, A_2, ..A_k$ is the set of partitions of the tree and $\lambda_1, \lambda_2, ..., \lambda_k$ the corresponding indexes, Let us define E in the following way: $E(x,y) = \max_{i=1,2,...,k} \{\lambda_i \text{ such that } (x,y) \in A_i\}$.

Example 5.8. Let $X = \{a_1, a_2, a_3, a_4, a_5\}$ and E the min-indistinguishability operator with matrix

$$\begin{pmatrix} 1 & 0.5 & 0.2 & 0.2 & 0.2 \\ 0.5 & 1 & 0.2 & 0.2 & 0.2 \\ 0.2 & 0.2 & 1 & 0.2 & 0.2 \\ 0.2 & 0.2 & 0.2 & 1 & 0.7 \\ 0.2 & 0.2 & 0.2 & 0.7 & 1 \end{pmatrix}.$$

The corresponding tree is

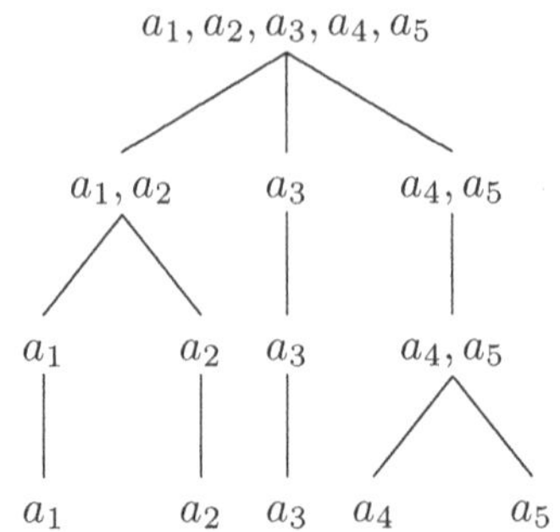

An important question with respect to indistinguishability operators is its storage, since in many situations, the cardinality of the set and consequently the dimension of its associated matrix can be very big. In this sense, in [131] a nice representation of min-indistinguishability operators is provided.

Proposition 5.9. *Let E be a* min-*indistinguishability operator on a finite set X. The elements of X can be reordered in such a way that in every file of the corresponding matrix the entries on the right of the diagonal are in non ascending order. Then the matrix is said to be in its normal form.*

The proof of this proposition is straightforward and can be found in [131].

Example 5.10. Let us consider the min-indistinguishability operator E on $X = \{x_1, x_2, x_3, x_4, x_5\}$ with matrix E

$$\begin{array}{c} \\ x_1 \\ x_2 \\ x_3 \\ x_4 \\ x_5 \end{array} \begin{array}{c} \begin{array}{ccccc} x_1 & x_2 & x_3 & x_4 & x_5 \end{array} \\ \begin{pmatrix} 1 & 0.3 & 0.7 & 0.2 & 0.8 \\ 0.3 & 1 & 0.3 & 0.2 & 0.3 \\ 0.7 & 0.3 & 1 & 0.2 & 0.7 \\ 0.2 & 0.2 & 0.2 & 1 & 0.2 \\ 0.8 & 0.3 & 0.7 & 0.2 & 1 \end{pmatrix} \end{array}.$$

After reordering the files and columns of E we obtain the matrix in its normal form E'

$$\begin{array}{c} \\ x_1 \\ x_5 \\ x_3 \\ x_2 \\ x_4 \end{array} \begin{array}{c} \begin{array}{ccccc} x_1 & x_5 & x_3 & x_2 & x_4 \end{array} \\ \begin{pmatrix} 1 & 0.8 & 0.7 & 0.3 & 0.2 \\ 0.8 & 1 & 0.7 & 0.3 & 0.2 \\ 0.7 & 0.7 & 1 & 0.3 & 0.2 \\ 0.3 & 0.3 & 0.3 & 1 & 0.2 \\ 0.2 & 0.2 & 0.2 & 0.2 & 1 \end{pmatrix} \end{array}.$$

Once the matrix is in its normal form, the sequence of the elements of X with the entries on the right of the diagonal of the matrix placed in between is called the representing sequence of the min-indistinguishability operator.

Example 5.11. The representing sequence of the preceding min-indistinguishability operator is

$$x_1 \; 0.8 \; x_5 \; 0.7 \; x_3 \; 0.3 \; x_2 \; 0.2 \; x_4.$$

Reciprocally, from a representing sequence

$$x_{\sigma(1)}\ s_1\ x_{\sigma(2)}\ s_2\ x_{\sigma(3)}\ \dots\ x_{\sigma(n-1)}\ s_{n-1}\ x_{\sigma(n)}$$

where σ is a permutation of X a min-indistinguishability operator $E = (a_{ij})$ on X can be generated it the following way:

$$E(x_{\sigma(i)}, x_{\sigma(j)}) = a_{\sigma(i)\sigma(j)} = \begin{cases} 1 & \text{if } i = j \\ \min(s_i, s_{i+1}, ..., s_j) & \text{if } \sigma(i) \leq \sigma(j) \\ \min(s_j, s_{j+1}, ..., s_i) & \text{if } \sigma(i) \geq \sigma(j). \end{cases}$$

The next Theorem summarizes these results.

Theorem 5.12. *There is a bijection between the set of* min-*indistinguishability operators on a set X and the set of representing sequences of X.*

Another interesting question about min-indistinguishability operators is how many essentially different such operators are there for a given n. Of course, the problem is equivalent to finding how many different hierarchical trees are there. It is a difficult combinatorial problem and the answer is only known for small values of n. The first one who studied it was Riera in 1978 ([119]) and in [50] a step forward has been made.

First of all, let us clarify what we understand as essentially different operators.

Lemma 5.13. *The (crisp) relation $\sim_1$ on the set of* min-*indistinguishability operators on X defined by $E \sim_1 F$ if and only if there exists an increasing map $f : [0,1] \to [0,1]$ such that $f(E(x,y)) = F(x,y)$ for all $x, y \in X$ is an equivalence relation.*

Proof. Trivial.

Lemma 5.14. *The (crisp) relation $\sim_2$ on the set of* min-*indistinguishability operators on X defined by $E \sim_2 F$ if and only if there exists a permutation $\sigma : X \to X$ such that $F(x,y) = E(\sigma(x), \sigma(y))$ is an equivalence relation.*

Proof. Trivial.

Combining the two equivalence relations $\sim_1$ and $\sim_2$ we obtain the next definition.

Definition 5.15. *Two* min-*indistinguishability operators E and F on a finite set X of cardinality n have the same structure if and only if there exists an increasing map $f : [0,1] \to [0,1]$ and a permutation $\sigma : X \to X$ such that $F(x,y) = f(E(\sigma(x), \sigma(y)))$.*

In order to determine the structures of min-indistinguishability operators the following Proposition is needed [88].

Proposition 5.16. *For any* min-*indistinguishability operator E on a finite set X of cardinality n there exists a decomposition $E = \sigma(E'(t; C_{n_1 \times n_1}, D_{n_2 \times n_2}))$ $(n_1 + n_2 = n)$ with*

$$E'(t; C_{n_1 \times n_1}, D_{n_2 \times n_2}) = \begin{pmatrix} C_{n_1 \times n_1} & (t)_{n_1 \times n_2} \\ (t)_{n_2 \times n_1} & D_{n_2 \times n_2} \end{pmatrix}.$$

where t is the smallest value of E, C and D are min-*indistinguishability operators and $\sigma : X \to X$ is a permutation.*

Example 5.17. The min-indistinguishability operator E

$$\begin{array}{c} \\ x_1 \\ x_2 \\ x_3 \\ x_4 \end{array} \begin{array}{c} \begin{array}{cccc} x_1 & x_2 & x_3 & x_4 \end{array} \\ \begin{pmatrix} 1 & 0.6 & 0.8 & 0.6 \\ 0.6 & 1 & 0.6 & 0.7 \\ 0.8 & 0.6 & 1 & 0.6 \\ 0.6 & 0.7 & 0.6 & 1 \end{pmatrix} \end{array}$$

can be decomposed into

$$C_{2\times 2} = \begin{pmatrix} 1 & 0.8 \\ 0.8 & 1 \end{pmatrix}$$

and

$$C_{2\times 2} = \begin{pmatrix} 1 & 0.7 \\ 0.7 & 1 \end{pmatrix}.$$

If σ is the permutation $(1, 3, 2, 4)$, then

$$E = \sigma(E'(0.6; C_{2\times 2}, D_{2\times 2})).$$

We will calculate all structures for $n \leq 5$ from the preceding decomposition.

$n = 2$.

There is only one structure

$$\begin{pmatrix} 1 & a \\ a & 1 \end{pmatrix}.$$

$n = 3$.

The decompositions with $(n_1 = 1,\, n_2 = 2)$ and $(n_1 = 2,\, n_2 = 1)$ represent the same structure with the permutation $\sigma = (3, 2, 1)$.

$$E'(t; C, D) = \begin{pmatrix} \begin{pmatrix} 1 & a \\ a & 1 \end{pmatrix} & (t)_{1\times 2} \\ (t)_{2\times 1} & 1 \end{pmatrix}$$

where $a \geq t$.

There are two possibilities

$$\begin{pmatrix} 1 & a & a \\ a & 1 & a \\ a & a & 1 \end{pmatrix}$$

and

$$\begin{pmatrix} 1 & a & t \\ a & 1 & t \\ t & t & 1 \end{pmatrix}.$$

with $a > t$.

$n = 4$.

The decompositions with ($n_1 = 1$, $n_2 = 3$) and ($n_1 = 3$, $n_2 = 1$) represent the same structure with the permutation $\sigma = (4, 2, 3, 1)$. The min-indistinguishability operators on sets of cardinality 4 are of two types.

- Decomposition of type ($n_1 = 3$, $n_2 = 1$):

$$\begin{pmatrix} \begin{pmatrix} 1 & a & b \\ a & 1 & b \\ b & b & a \end{pmatrix} & (t)_{1\times 3} \\ (t)_{3\times 1} & 1 \end{pmatrix}$$

  with $a \leq b \leq t$. Replacing each inequality by a strict one or by an equality there are 4 possible structures.

- Decomposition with ($n_1 = 2$, $n_2 = 2$):

$$\begin{pmatrix} \begin{pmatrix} 1 & a \\ a & 1 \end{pmatrix} & (t)_{2\times 2} \\ (t)_{2\times 2} & \begin{pmatrix} 1 & a \\ a & 1 \end{pmatrix} \end{pmatrix}$$

with $a \geq t$, $b \geq t$. The cases with $b = t$ have already been counted in the preceding decomposition. This gives 2 more structures.

There are therefore 6 different structures for $n = 4$.

$n = 5$.

- The decompositions with ($n_1 = 1$, $n_2 = 4$) and ($n_1 = 4$, $n_2 = 1$) represent the same structure with the permutation $\sigma = (5, 2, 3, 4, 1)$.
- The decompositions with ($n_1 = 2$, $n_2 = 3$): Let $a \in C$ and $b, c \in D$. Then

$$a \geq t$$
$$b \geq c \geq t.$$

Possible combinations of the values a, b, c:

1. $a \geq b \geq c \geq t$. 2^3 different cases.
2. $b \geq a \geq c \geq t$. 2^2 different cases. (The case $a = b$ had already been counted).

3. $b \geq c \geq a \geq t$. 2 more additional cases. (The case $a = c$ has already been counted and $a = t$ has been counted in the preceding decomposition).

Summing up, the number of structures for $n = 5$ is 22.

The next table 5.1 summarizes these results.

Table 5.1 Number of structures

n	Number of structures
1	1
2	1
3	2
4	6
5	22

Let us find the trees associated to each structure.

The next results are straightforward.

Proposition 5.18. *Two* min*-indistinguishability operators E and F generate the same non-indexed tree if and only if $E \sim_1 F$.*

Proposition 5.19. *Two* min*-indistinguishability operators E and F generate the same indexed tree except for permutations of the branches if and only if $E \sim_2 F$.*

Proposition 5.20. *Two* min*-indistinguishability operators E and F generate the same non-indexed tree except for permutations of the branches if and only if they are the same structure.*

Therefore searching for the different structures is equivalent to searching the different non-indexed trees except for permutations of their branches.

The tables 5.2 and 5.3 show the essentially different hierarchical trees for $n = 3$ and $n = 4$.

Table 5.2 The 2 hierarchical trees for $n = 3$

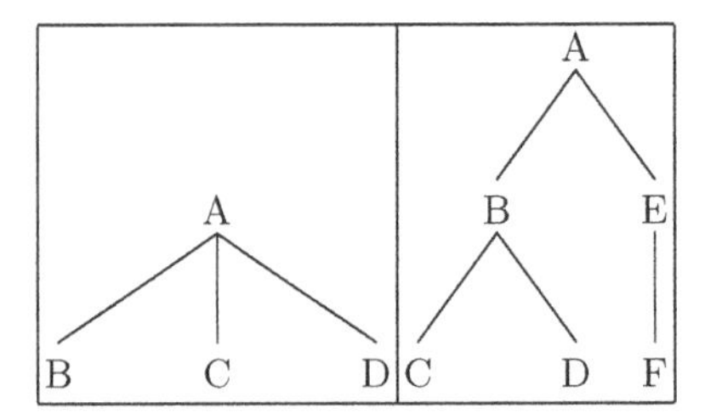

Table 5.3 The 6 hierarchical trees for $n = 4$

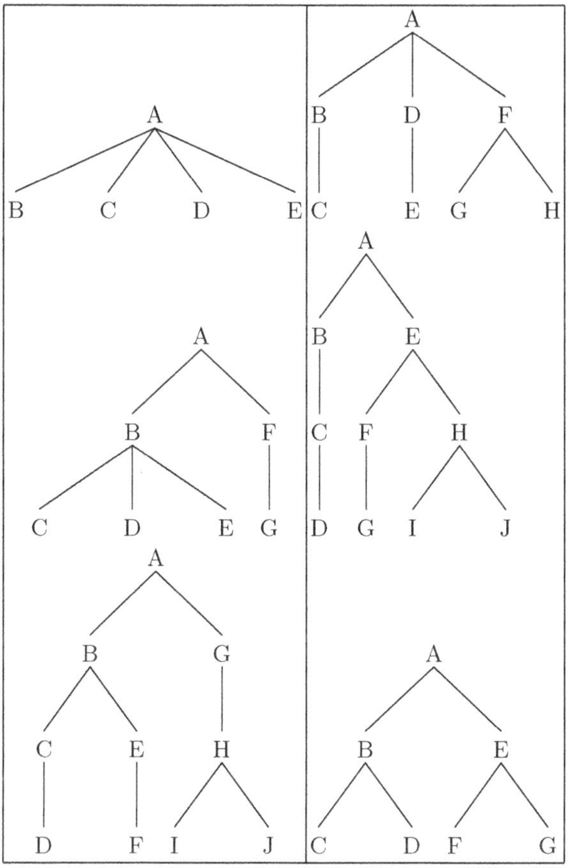

6

Betweenness Relations

As explained in Chapter 4, T-indistinguishability operators have a very important metric component. One consequence of this fact is that, if the t-norm is continuous Archimedean, the operators generate metric betweenness relations and their structure can be studied in terms of the different types of betweenness relations they produce. This chapter will revisit the three main methods for building T-indistinguishability operators -with the sup $-T$ product, using the Representation Theorem and by constructing a decomposable relation from a fuzzy subset- in relation to betweenness relations.

The structure of the betweenness relations generated by a T-indistinguishability operator depends on the length (Definition 6.16 and dimension of the operator. Since the length and dimension of a T-indistinguishability operator are related to its generation via sup $-T$ product and via the Representation Theorem, respectively, betweenness relations are an ideal tool for linking the two methods. For instance, in Section 6.5, they are used to prove the equivalence between one dimensionality and maximality of the length. This result can be generalized in the following rule of thumb:

The greater the dimension the smaller the length, and vice versa.

The betweenness relations generated by decomposable indistinguishability operators are very special. They are called radial, meaning that there exists exactly one element in between any other two. This kind of relations is also characteristic of decomposable indistinguishability operators.

When a set has some additional structure, such as an operation or an ordering relation, it seems reasonable to ask for a generated indistinguishability operators to be compatible with it. Section 6.4 will discuss this issue for indistinguishability operators generated by fuzzy numbers on the real line.

It may happen that, due to some perturbations in the values of an indistinguishability operator, some betweenness relations that would appear in the original one are not present in the new distorted operator. To handle this problem fuzzy betweenness relations are introduced that make it possible to say that an element is more or less between two others.

J. Recasens: Indistinguishability Operator, STUDFUZZ 260, pp. 107–124.
springerlink.com

6.1 Betweenness Relations

The notion of metric betweenness relation appears in [94] where it is defined as follows:

Definition 6.1. *A (metric) betweenness relation on a set X is a ternary relation B on X (i.e. $B \subseteq X^3$) satisfying for all $x, y, z \in X$*

1. $(x, y, z) \in B \Rightarrow x \neq y \neq z \neq x$
2. $(x, y, z) \in B \Rightarrow (z, y, x) \in B$
3. $(x, y, z) \in B \Rightarrow (y, z, x) \notin B, (z, x, y) \notin B$
4. $(x, y, z) \in B$ *and* $(x, z, t) \in B \Rightarrow (x, y, t) \in B$ *and* $(y, z, t) \in B$.

If $(x, y, z) \in B$, then y is said to be between x and z.

If given any three elements of B, one of them is between the other two, then the betweenness relation is called linear or total.

The idea of metric betweenness appeared in the study of metric spaces. If d is a distance defined on a set X, the relation "y is between x and z when $d(x, y) + d(y, z) = d(x, z)$" satisfies the axioms of a betweenness relation and Menger used these relations in the study of isometric embeddings of metric spaces.

The fact that indistinguishability operators separating points define betweenness relations (when T is a continuous Archimedean t-norm) provides them with a metric flavour that facilitates their study from a metric point of view.

Proposition 6.2. *Let T be a continuous Archimedean t-norm and E a T-indistinguishability operator separating points on a set X such that $E(x, y) \neq 0$ for all $x, y \in X$. The ternary relation B on X defined by $(x, y, z) \in B$ if and only if $x \neq y \neq z \neq x$ and*

$$T(E(x, y), E(y, z)) = E(x, z)$$

is a betweenness relation on X.

Proof

1. Trivial.
2. follows from the commutativity of T and the symmetry of E.
3. Let us prove for instance that if $(x, y, z) \in B$, then $(y, z, x) \notin B$.
   If $T(E(y, z), E(z, x)) = E(y, x)$, then

$$T(E(y, z), E(z, x), E(y, z)) = T(E(y, x), E(y, z)) = E(x, z)$$

   and therefore

$$T(E(y, z), E(y, z), E(x, z)) = E(x, z).$$

   Since $E(x, z) \neq 0$, this implies that $T(E(y, z), E(y, z)) = 1$ and also $E(y, z) = 1$ contradicting the separability of E.

4. Let us prove, for instance, that if $(x, y, z) \in B$ and $(x, z, t) \in B$, then $(y, z, t) \in B$.

$$\begin{aligned} E(x,t) &= T(E(x,z), E(z,t)) \\ &= T(T(E(x,y), E(y,z)), E(z,t)) \\ &= T(E(x,y), T(E(y,z), E(z,t))) \end{aligned}$$

and since

$$T(E(y,z), E(z,t)) \leq E(y,t),$$

monotonicity of T assures that

$$\begin{aligned} E(x,t) &\geq T(E(x,y), E(y,t)) \\ &\geq T(E(x,y), T(E(y,z), E(z,t))) = E(x,t) \end{aligned}$$

and therefore

$$T(E(x,y), E(y,t)) = T(E(x,y), (E(y,z), E(z,t)) \neq 0.$$

Since $E(x,y) \neq 0$,

$$E(y,t) = T(E(y,z), E(z,t)).$$

6.2 Linear Betweenness Relations and One Dimensional Indistinguishability Operators

The structure of the betweenness relation generated by a T-indistinguishability operator reflects its combinatorial complexity expressed by its dimension.

Proposition 6.3. *Let T be a continuous Archimedean t-norm and E a T-indistinguishability operator separating points on X such that there exists* $\min\{E(x,y) \mid x, y \in X\} \neq 0$ *for all $x, y \in X$. E is one dimensional if and only if the betweenness relation B determined by E on X is linear.*

Proof

$\Rightarrow$)

Let μ be a fuzzy set of X generating E (i.e. $E = E_\mu$). Since E separates points, μ is a one to one map and defines a total ordering on X.

$$x \leq_\mu y \text{ if and only if } \mu(x) \leq \mu(y).$$

If $x <_\mu y <_\mu z$, then

$$\begin{aligned} &T(E(x,y), E(y,z)) \\ &= t^{[-1]}(t(t^{-1}(t(\mu(x)) - t(\mu(y))) + t(t^{-1}(t(\mu(y)) - t(\mu(z)))) \\ &= t^{[-1]}(t(\mu(x)) - t(\mu(z))) = E(x,z). \end{aligned}$$

$\Leftarrow$) Let $a, b \in X$ be such that $E(a,b) = \min\{E(x,y) \mid x, y \in X\}$ and consider the column μ_a of E.

If $x, y \in X$ satisfy (a, x, y), then

$$E(a,y) = T(E(a,x), E(x,y))$$

and

$$E(x,y) = \overrightarrow{T}(E(a,x)|E(a,y)) = \overrightarrow{T}(\mu_a(x)|\mu_a(y))$$

and μ_a is a generator of E.

It is worth noting that in [13] a characterization theorem of one-dimensional T-indistinguishability operators for general continuous t-norms is proved that generalizes the previous result.

Corollary 6.4. *Let T be a continuous Archimedean t-norm and E a T-indistinguishability operator separating points on a finite set X of cardinality n satisfying $E(x,y) \neq 0$ $\forall x, y \in X$. E is one dimensional if and only if the cardinality of B is $2 \cdot \binom{n}{3}$.*

Proof. Trivial. The factor 2 is due to the fact that if $(x,y,z) \in B$, then $(z,y,x) \in B$.

The tight link between the betweenness relation defined by a T-indistinguishability operator and its dimension is also shown in the next proposition.

Proposition 6.5. [116] *Let T be a continuous Archimedean t-norm and E be a T-indistinguishability operator separating points on a finite set X of cardinality n satisfying $E(x,y) \neq 0$ $\forall x, y \in X$. If E is bidimensional, then the cardinality of B is greater than or equal to $2 \cdot T(n,5,3)$ where $T(n,5,3)$ is a Turán number that is conjectured to be $2 \cdot \binom{\frac{n}{2}}{3}$ if n is even and $\binom{\frac{n+1}{2}}{3} + \binom{\frac{n-1}{2}}{3}$ if n is odd* [116] [43].

6.3 Radial Betweenness Relations and Decomposable Indistinguishability Operators

Apart from linear betweenness relations, there is another simple kind of such relations that we call radial: There is exactly a central element that is between any other two. This is a very common structure that reflects the behaviour of centralized systems where there is a central kernel that controls all information and relations in the system. It turns out that these radial relations characterize decomposable indistinguishability operators.

Definition 6.6. *A betweenness relation B on a set X is called radial if and only if there exists an element $a \in X$ such that a is between any other two elements of X and these are the only elements of B (i.e. $(x,y,z) \in B$ if and only if $y = a$). The element a is called the center of the betweenness relation.*

Proposition 6.7. *Let T be a continuous Archimedean t-norm, E a T-indistinguishability operator separating points on a finite set X of cardinality n satisfying $E(x,y) \neq 0$ $\forall x, y \in X$ and B the betweenness relation generated on X by E. E is decomposable if and only if B is radial or E can be extended to a T-indistinguishability operator $\overline{E}$ on $\overline{X} = X \cup \{a\}$ with $a \notin X$ in such a way that the betweenness relation $\overline{B}$ generated on $\overline{X}$ by $\overline{E}$ is radial with center a.*

Proof

$\Rightarrow$)

Let $E(x,y) = T(\mu(x), \mu(y))$ if $x \neq y$.

- If there exists $x_0 \in X$ such that $\mu(x_0) = 1$, then we are in the first case, since
  - given $x, y \in X$ with $x \neq y \neq x_0 \neq x$,

$$\begin{aligned} T(E(x,x_0), E(x_0,y)) &= T(T(\mu(x), \mu(x_0)), T(\mu(x_0), \mu(y))) \\ &= T(\mu(x), \mu(y)) = E(x,y) \end{aligned}$$

    and therefore x_0 is between any other two elements of X.
  - If $z \neq x_0$ and $x \neq z \neq y \neq x$, then

$$\begin{aligned} T(E(x,z), E(z,y)) &= T(T(\mu(x), \mu(z)), T(\mu(z), \mu(y))) \\ &= T(T(\mu(z), \mu(z)), T(\mu(x), \mu(y))) \\ &= T(T(\mu(z), \mu(z)), E(x,y)) < E(x,y). \end{aligned}$$

    Therefore, z is not between x and y.
- If there exists no $x_0 \in X$ with $\mu(x_0) = 1$, then we can define the fuzzy subset $\overline{\mu}$ of $\overline{X}$ by $\overline{\mu}(x) = \mu(x)$ $\forall x \in X$ and $\overline{\mu}(a) = 1$ and consider the relation $\overline{E}$ on $\overline{X}$ generated by $\overline{\mu}$.

  $\Leftarrow$)

  If B is radial with center c, then the column μ_c of E generates E.

  Indeed: if $x \neq c \neq y \neq x$, then

$$E(x,y) = T(E(x,c), E(c,y)) = T(\mu_c(x), \mu_c(y)).$$

In the second case, E is generated by the fuzzy subset μ of X defined by $\mu(x) = \overline{E}(x,a)$ $\forall x \in X$.

6.4 Fuzzy Numbers and Betweenness Relations

From a fuzzy subset μ of a given universe the one dimensional indistinguishability operator E_μ and the decomposable operator E^μ can be generated, but in both cases, we do not assume or take any structure on our universe into account. Nevertheless, it is common in fuzzy systems to deal with rich structures such as subsets of the real line, $\mathbb{R}^n$ or, more general, ordered sets. In these cases we should be able to use the underlying structure while generating indistinguishability operators.

In this section we give a way to obtain indistinguishability operators from fuzzy numbers that generate betweenness relations compatible with the ordering of the real line. This can be applied to lattices and posets as well. The obtained operators are between E_μ and E^μ.

Lemma 6.8. $E_\mu \geq E^\mu$.

Proof. It is a consequence of $E_T(x,y) \geq T(x,y)$.

Lemma 6.9. *Let μ be a fuzzy subset of X and E_μ the T-indistinguishability operator generated by μ. If $\mu(x) \leq \mu(y)$ then*

$$T(E_\mu(x,y), \mu(y)) = \mu(x).$$

Proof. If $\mu(x) \leq \mu(y)$ then

$$E_\mu(x,y) = \overrightarrow{T}(\mu(y)|\mu(x)).$$

Let us suppose that our fuzzy subset μ is normal with $\mu(a) = 1$ and a one to one map. Then $E_\mu(x,a) = E_T(\mu(x), \mu(a)) = E_T(\mu(x), 1) = \mu(x)$ for all $x \in X$ and the betweenness relation B_μ generated by E_μ is such that a is never between any couple of elements of X. Moreover the betweenness relation is linear compatible with the order in $[0,1]$ in the sense that if $\mu(x) \leq \mu(y) \leq \mu(z)$ then $(x,y,z) \in B_\mu$.

Definition 6.10. *A fuzzy number is a map $\mu_a : \mathbb{R} \to [0,1]$ such that there exists $a \in \mathbb{R}$ with $\mu_a(a) = 1$, non decreasing in $(-\infty, a)$ and non increasing in $(a, +\infty)$.*

Being a fuzzy number μ_a a fuzzy subset of $\mathbb{R}$, it generates two indistinguishability operators E_μ and E^μ on $\mathbb{R}$.

Nevertheless, it seems reasonable to impose a kind of compatibility with the ordering of the real line and the betweenness relation generated by the indistinguishability operator. This can be achieved with the following definition.

Definition 6.11. *Let μ_a be a fuzzy number. The fuzzy relation E_a associated to μ_a is defined by*

$$E_a(x,y) = \begin{cases} E_{\mu_a}(x,y) & \text{if } x,y \leq a \text{ or } x,y \geq a \\ E^{\mu_a}(x,y) & \text{otherwise.} \end{cases}$$

Proposition 6.12. *The fuzzy relation E_a associated to a fuzzy number μ_a is a T-indistinguishability operator.*

Proof. Reflexivity and symmetry are trivial.
Transitivity:

a) If $E_a(x,y) = E_{\mu_a}(x,y)$ and $E_a(y,z) = E_{\mu_a}(y,z)$, then $E_a(x,z) = E_{\mu_a}(x,z)$ and transitivity follows.
b) If $E_a(x,y) = E^{\mu_a}(x,y)$ and $E_a(y,z) = E^{\mu_a}(y,z)$, then $E_a(x,z) = E_{\mu_a}(x,z)$. Since $E^{\mu_a}(x,y) \leq E_{\mu_a}(x,y)$,

$$T(E_a(x,y), E_a(y,z)) \leq T(E_{\mu_a}(x,y), E_{\mu_a}(y,z)) \leq E_{\mu_a}(x,z).$$

c) If $E_a(x,y) = E_{\mu_a}(x,y)$ and $E_a(y,z) = E^{\mu_a}(y,z)$ then $E_a(x,z) = E^{\mu_a}(x,z)$.

$$\begin{aligned} T(E_a(x,y), E_a(y,z)) &= T(E_{\mu_a}(x,y), E^{\mu_a}(y,z)) \\ &= T(E_{\mu_a}(x,y), \mu_a(y), \mu_a(z)) \end{aligned} \quad (1)$$

If $\mu(x) \leq \mu(y)$ then, using Lemma 6.9,
$(1) = T(\mu_a(x), \mu_a(z)) = E^{\mu_a}(x,z)$.
If $\mu(y) \leq \mu(x)$ then, using Lemma 6.9,
$(1) = T(\mu_a(y), \mu_a(z)) \leq T(\mu_a(x), \mu_a(z)) = E^{\mu_a}(x,z)$.

Proposition 6.13. *Let T be a continuous Archimedean t-norm. The T-indistinguishability operator E_a obtained from the fuzzy number μ_a is compatible with the ordering of the real line in the sense that if $x \leq y \leq z$, then*

$$T(E_a(x,y), E_a(y,z)) = E_a(x,z).$$

Proof. If $x \leq y \leq z \leq a$ or $a \leq x \leq y \leq z$, it follows easily from the linearity of the betweenness relation.

If $x \leq y \leq a \leq z$ or $x \leq a \leq y \leq z$, it follows easily from the radial betweenness relation.

Reciprocally,

Proposition 6.14. *Let T be a continuous Archimedean t-norm and E_a the T-indistinguishability operator obtained from the fuzzy number μ_a. If*

$$T(E_a(x,y), E_a(y,z)) = E_a(x,z),$$

then

$$x \leq y \leq z \text{ or } z \leq y \leq x.$$

Example 6.15. Let μ_3 be the triangular fuzzy number [2,3,4] (Figure 6.1) and T the Łukasiewicz t-norm. The T-indistinguishability operators E_{μ_3}, E^{μ_3} and E_3 associated to μ_3 are defined for all $x, y \in \mathbb{R}$ by

$$E_{\mu_3}(x,y) = 1 - |\mu_3(x) - \mu_3(y)|$$

$$E^{\mu_3}(x,y) = \begin{cases} \max(\mu_3(x) + \mu_3(y) - 1, 0) & \text{if } x \neq y \\ 1 & \text{otherwise} \end{cases}$$

$$E_3(x,y) = \begin{cases} 1 - |\mu_3(x) - \mu_3(y)| & \text{if } x, y \leq 3 \text{ or } x, y \geq 3 \\ \max(\mu_3(x) + \mu_3(y) - 1, 0) & \text{otherwise.} \end{cases}$$

(See Figures 6.2, 6.3 and 6.4).

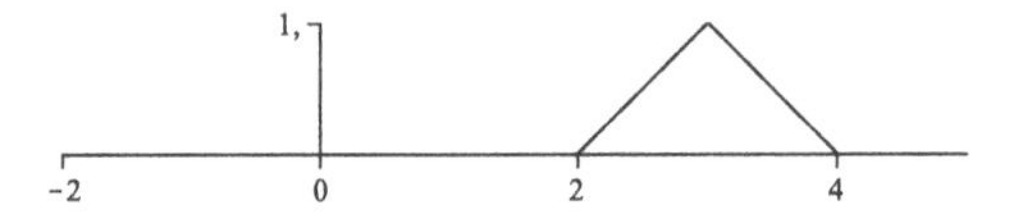

Fig. 6.1 μ_3

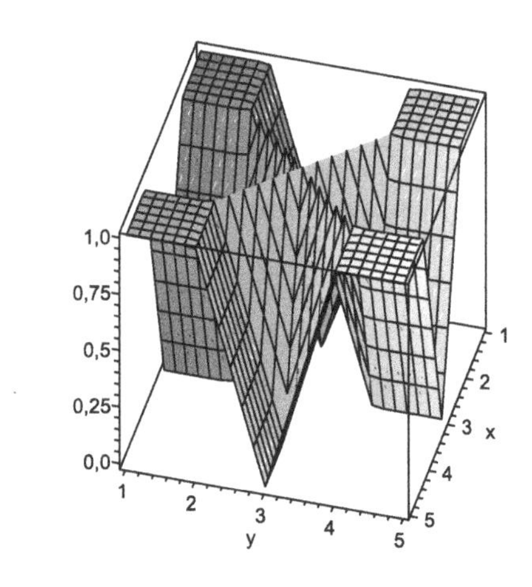

Fig. 6.2 E_{μ_3}

It is interesting to note that with triangular and trapezoidal numbers, the T-indistinguishability operator obtained when T is the Łukasiewicz t-norm is of dimension 2. For example, the previous E_3 is the infimum of the T-indistinguishability operators E_{μ_3} and E_μ where μ is the fuzzy subset of the real line defined by

$$\mu(x) = \begin{cases} 0 \text{ if } x \leq 2.5 \\ x - 2.5 \text{ if } 2.5 \leq x \leq 3.5 \\ 1 \text{ if } 3.5 \leq x \end{cases}$$

μ and E_μ are displayed in Figures 6.5 and 6.6.

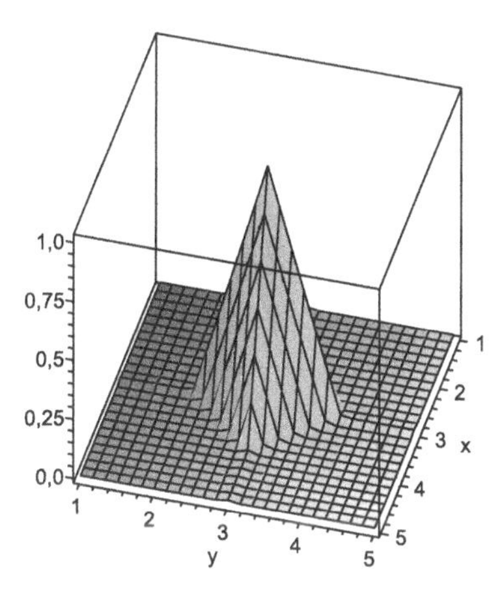

Fig. 6.3 E^{μ_3}

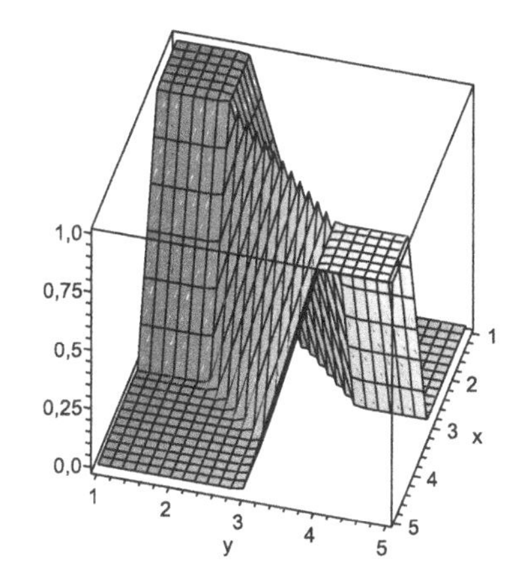

Fig. 6.4 E_3

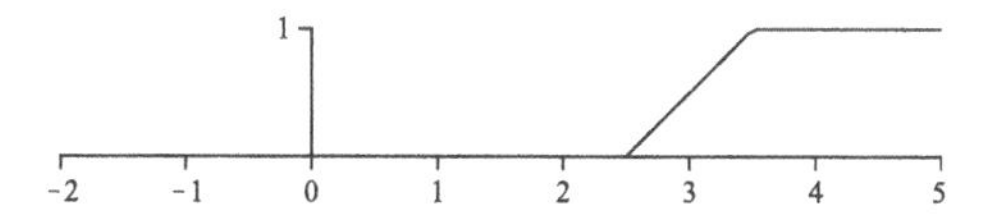

Fig. 6.5 μ

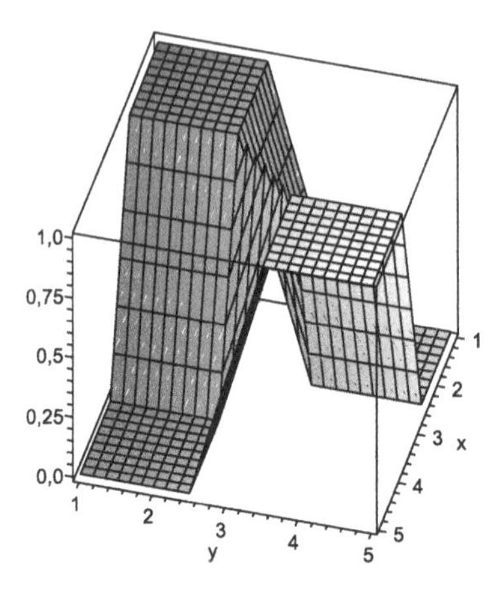

Fig. 6.6 E_μ

6.5 The Length of an Indistinguishability Operator and Betweenness Relations

Let us now introduce the notion of length of a T-indistinguishability operator and study its relation with its dimension and decomposability.

Definition 6.16. *Given a t-norm T and a T-indistinguishability operator E on a set X , the length of E is the maximum $k \in \mathbb{N}$ (if it exists) such that there exists a reflexive and symmetric fuzzy relation R on X with $R^{k-1} \neq R^k = E$ and* $\text{length}(E) = \infty$ *otherwise.*

Note that $\text{length}(E) \geq 1$, since $E^1 = E$, and if X is finite of cardinality n, then $\text{length}(E) \leq n-1$.

Lemma 6.17. *Let R be a fuzzy relation on a finite set X, $a, b \in X$, T a continuous t-norm and k a positive integer. Then*

$$R^k(a,b) = \sup_{x_2,\ldots,x_k \in X} T(R(a,x_2), R(x_2,x_3), \ldots, R(x_k,b)).$$

Proof. Let us prove the assertion by induction.

The assertion is trivially true for $k = 1, 2$. Suppose the result true for $k = n$.

$$\begin{aligned} R^{n+1}(a,b) &= (R^n \circ R)(a,b) \\ &= \sup_{x_{n+1} \in X} T(R^n(a,x_{n+1}), R(x_{n+1},b)) \\ &= \sup_{x_{n+1} \in X} T(\sup_{x_2,\ldots,x_n \in X} T(R(a,x_2),\ldots,R(x_n,x_{n+1})), R(x_{n+1},b)) \\ &= \sup_{x_2,\ldots,x_{n+1} \in X} T(R(a,x_2),\ldots,R(x_{n+1},b)). \end{aligned}$$

The next lemma will be the cornerstone to relate length with dimension and decomposability.

From now until the end of this section, T will be a continuous Archimedean t-norm and E a T-indistinguishability operator on X separating points such that $E(x,y) \neq 0 \ \forall x, y \in X$.

Lemma 6.18. *Let E be a T-indistinguishability operator on a finite set X and B the betweenness relation defined by E. If* length$(E) = l$*, then there exist $x_1, x_2, ..., x_{l+1} \in X$ such that $(x_i, x_j, x_k) \in B$ if $i < j < k$.*

Proof. If length$(E) = l$, then there exists a reflexive and symmetric fuzzy relation R on X such that $E = R^l > R^{l-1}$. Therefore, there exist $x_1, x_{l+1} \in X$ such that $R^l(x_1, x_{l+1}) > R^{l-1}(x_1, x_{l+1})$. On the other hand, by Lemma 6.17,

$$R^l(x_1, x_{l+1}) = T(R(x_1, x_2), R(x_2, x_3), ..., R(x_l, x_{l+1})) \tag{1}$$

for some $x_2, x_3, ..., x_l \in X$.

Moreover,

$$R^{j-i}(x_i, x_j) = T(R(x_i, x_{i+1}), R(x_{i+1}, x_{i+2}), ..., R(x_{j-1}, x_j)) \text{ for all } j > i, \tag{2}$$

since

$$\begin{aligned} R^{j-i}(x_i, x_j) &= T(R(x_i, y_{i+1}), R(y_{i+1}, y_{i+2}), ..., R(y_{j-1}, x_j)) \\ &> T(R(x_i, x_{i+1}), R(x_{i+1}, x_{i+2}), ..., R(x_{j-1}, x_j)) \end{aligned}$$

for some $y_{i+1}, ..., y_{j-1} \in X$ would contradict (1).

From this fact it follows that

$$E(x_i, x_j) = R^{j-i}(x_i, x_j) \ \forall j > i, \tag{3}$$

since, if this were false, there would exist $k > j - i$ such that

$$E(x_i, x_j) = R^k(x_i, x_j) > R^{j-i}(x_i, x_j)$$

and therefore,

$$\begin{aligned} R^k(x_i, x_j) &= T(R(x_i, y_2), R(y_2, y_3), ..., R(y_k, x_j)) \\ &> T(R(x_i, x_{i+1}), ..., R(x_{j-1}, x_j)) \end{aligned}$$

for some $y_2, y_3, ..., y_k \in X$.

Replacing

$$R(x_i, x_{i+1}), ..., R(x_{j-1}, x_j)$$

by

$$R(x_i, y_2), R(y_2, y_3), ..., R(y_k, x_j)$$

in (1) we would have

$$R^m(x_1, x_{l+1}) > R^l(x_1, x_{l+1})$$

contradicting the maximality of l.

Now, from (2) and (3) the result follows easily:

If $i < j < k$, then

$$\begin{aligned} &T(E(x_i, x_j), E(x_j, x_k)) \\ &= T(R(x_i, x_{i+1}), ..., R(x_{j-1}, x_j), R(x_j, x_{j+1}), ..., R(x_{k-1}, x_k)) \\ &= R^{k-i}(x_i, x_k) = E(x_i, x_k) \end{aligned}$$

and therefore $(x_i, x_j, x_k) \in B$.

Lemma 6.19. *Given a continuous Archimedean t-norm T, t an additive generator of T, $m \in \mathbb{N}$ and $a \in]0,1]$, there exists $\in [0,1[$ such that $T(a, b, \overset{m}{\ldots}, b) \neq 0$.*

Proof. For any $x \in [0,1]$, $T(a, x, \overset{m}{\ldots}, x) = t^{[-1]}(t(a) + mt(x))$.

If $mt(x) < t(0) - t(a)$, then $T(a, x, \overset{m}{\ldots}, x) \neq 0$.

Therefore, taking $b > t^{-1}(\frac{1-t(a)}{m})$ the lemma follows.

As a consequence of these two lemmas we can prove the following results.

Proposition 6.20. *Let E b a T-indistinguishability operator on a finite set X of cardinality n. E is one dimensional if and only if* length$(E) = n - 1$.

Proof

$\Rightarrow$) Let us order the elements $x_1, x_2, ..., x_n \in X$ in such a way that $(x_i, x_j, x_k) \in B$ if and only if $i < j < k$ or $k < j < i$, where B denotes the total betweenness relation on X generated by E.

Let $a = E(x_1, x_n)$. Due to Lemma 6.19, there exists $b \in [0,1[$ such that $T(a, b, \overset{n-2}{\ldots}, b) \neq 0$.

Let us define a reflexive and symmetric fuzzy relation R on X by

$$R(x_i, x_j) = \begin{cases} 1 & \text{if } i = j \\ E(x_i, x_j) & \text{if } |i - j| = 1 \\ T(E(x_i, x_j), b, \overset{k-1}{\ldots}, b) & \text{if } |i - j| = k > 1 \end{cases}$$

R satisfies $R^{n-2} \neq R^{n-1} = E$.

$\Leftarrow$) From Lemma 6.18, the betweenness relation defined by E is linear and therefore E is one dimensional.

Proposition 6.21. *Let E be a T-indistinguishability operator on a finite set X of cardinality n and* length$(E) = n - 2$. *Then E is bidimensional.*

Proof. From Lemma 6.18, all the elements of X but one ($a \in X$) form a chain $x_1, x_2, ..., x_n \in X$ such that $(x_i, x_j, x_k) \in B$ if $i < j < k$.

Let us prove that the fuzzy subsets columns μ_{x_1} and μ_a generate E: Given $x_i, x_j \in X$ with $i, j = 1, 2, ..., n-1$ and $i < j$

$$T(E(x_1, x_i), E(x_i, x_j)) = E(x_1, x_j)$$

or

$$T(\mu_{x_1}(x_i), E(x_i, x_j)) = \mu_{x_1}(x_j).$$

Therefore, since $\mu_{x_1}(x_i) > \mu_{x_1}(x_j)$,

$$E(x_i, x_j) = \overrightarrow{T}(\mu_{x_1}(x_i)|\mu_{x_1}(x_j)) = E_T(\mu_{x_1}(x_i), \mu_{x_1}(x_j)).$$

On the other hand, given $x_i \in X$ $i = 1, 2, ..., n-1$,

$$E(a, x_i) = \mu_a(x_i) = E_T(\mu_a(a), \mu_a(x_i)).$$

So, the dimension of E is less or equal than 2. If E were one dimensional, then the length of E would be $n-1$ from the preceding theorem. Therefore, the dimension of E is 2.

In the same way, the following result can be proved.

Proposition 6.22. *Let E be a T-indistinguishability operator on a finite set X of cardinality n and* length$(E) = k$. *Then the dimension of E is less or equal than $n - k$.*

Proposition 6.23. *Let E be a T-indistinguishability operator on a finite set X and B the betweenness relation defined by E on X. Then,* length$(E) = 1$ *if and only if $B = \emptyset$.*

Proof. Lemma 6.18.

Proposition 6.24. *If E is a decomposable T-indistinguishability operator on a finite set X, then* length$(E) \leq 2$.

Proof. The betweenness relation B generated by E is empty or radial. If B is empty, then length$(E) = 1$ by Proposition 6.23. If B is radial, then by Lemma 6.18 length$(E) \leq 2$.

It is interesting to note that these results cannot be generalized to non-Archimedean t-norms, since in this case indistinguishability operators do not define betweenness relations. For instance,

$$\begin{pmatrix} 1 & 0.5 & 0.2 \\ 0.5 & 1 & 0.2 \\ 0.2 & 0.2 & 1 \end{pmatrix}$$

and

$$\begin{pmatrix} 1 & 0.2 & 0.2 \\ 0.2 & 1 & 0.2 \\ 0.2 & 0.2 & 1 \end{pmatrix}$$

are two min-indistinguishability operators of length 2 and the first one is one dimensional while the second one has dimension two.

6.6 Fuzzy Betweenness Relations

Let us suppose that the values of a one-dimensional T-indistinguishability operator are distorted by some noise. Then the betweenness relation generated by this new relation will probably be not linear and even can be empty. This means that the definition of betweenness relation can not capture the possibility of a relation to be "almost" linear or -more generally speaking- is not capable of dealing with points being "more or less" between others.

In this section we fuzzify the definition of betweenness relation in order to handle these situations. It will be proved that fuzzy betweenness relations can be generated by T-indistinguishability operators separating points and, reciprocally, that every fuzzy betweenness relation generates such a T-indistinguishability operator.

In this section T will be a continuous strict Archimedean t-norm.

Definition 6.25. *A fuzzy betweenness relation with respect to a given strict Archimedean t-norm T on a set X is a fuzzy ternary relation, i.e. a map*

$$\begin{aligned} X \times X \times X &\to [0,1] \\ (x,y,z) &\to xyz \end{aligned}$$

satisfying the following properties for all $x,y,z,t \in X$

1. $xxy = 1$
2. $xyz = zyx$
3. a. $T(xyz, xzt) \leq xyt$
   b. $T(xyz, xzt) \leq yzt$
4. *If $x \neq y$, then $xyx < 1$.*

Lemma 6.26

1. *If $xyz = 1$ and $y \neq z$, then $xzy < 1$.*
2. *If $xyz = 1$ and $x \neq y$, then $zxy < 1$.*

Proof. *From 6.25.3.b) and 6.25.4., it follows $T(xyz, xzy) \leq yzy < 1$ and therefore $xzy < 1$.*

In a similar way, we can prove 2.

Corollary 6.27. *If $xyz = 1$ and $x \neq y \neq z \neq x$, then $xzy < 1$ and $zxy < 1$.*

Corollary 6.28. *The crisp part of a fuzzy betweenness relation in the set of triplets of different elements of X is a classical betweenness relation on X.*

Proposition 6.29. *Let T be a continuous strict Archimedean t-norm and E a T-indistinguishability operator separating points on X with $E(x,y) \neq 0$ for all $x, y \in X$. Then the fuzzy ternary relation on X defined by*

$$xyz = \overrightarrow{T}(E(x,z)|T(E(x,y),E(y,z)))$$

is a fuzzy betweenness relation.

Proof

- 6.25.1.

$$\begin{aligned} xxy &= \overrightarrow{T}(E(x,y)|T(E(x,x),E(x,y))) \\ &= \overrightarrow{T}(E(x,y)|E(x,y)) = 1. \end{aligned}$$

- 6.25.2. is immediate.
- 6.25.3.a) Applying the preceding definition to 6.25.3.a) we obtain the following inequality

$$\begin{aligned} T(\overrightarrow{T}(E(x,z)|T(E(x,y),E(y,z))),\overrightarrow{T}(E(x,t)|T(E(x,z),E(z,t)))) \\ \leq \overrightarrow{T}(E(x,t)|T(E(x,y),E(y,t))). \end{aligned}$$

In order to prove this inequality we can express it in terms of an additive generator t of T. Then the mentioned inequality can be written as

$$\begin{aligned} &t^{-1}(t(\overrightarrow{T}(E(x,z)|T(E(x,y),E(y,z)))) \\ &+t(\overrightarrow{T}(E(x,t)|T(E(x,z),E(z,t))))) \\ &\leq \overrightarrow{T}(E(x,t)|T(E(x,y),E(y,t))). \end{aligned}$$

If we represent the t-norms and its residuation using t, applying t to both sides of the inequality we obtain

$$\begin{aligned} t(E(x,y)) + t(E(y,z)) - t(E(x,z)) + t(E(x,z)) + t(E(z,t)) - t(E(x,t)) \\ \geq t(E(x,y)) + t(E(y,t)) - t(E(x,t)). \end{aligned}$$

Simplifying and applying t^{-1} to both sides, we finally obtain

$$t^{-1}(t(E(y,z)) + t(E(z,t))) \leq E(y,t).$$

The last inequality expresses the T-transitivity of E.
- 6.25.3.b) can be proved in a similar way as 6.25.3.a).
- 6.25.4. Since $E(x,y) \neq 0$ for all $x, y \in X$, for $x \neq y$ we have

$$\begin{aligned} xyx &= \overrightarrow{T}(E(x,x)|T(E(x,y),E(x,y))) \\ &= T(E(x,y),E(x,y)) < 1. \end{aligned}$$

It is worth noticing that given a fuzzy betweenness relation defined by means of a T-indistinguishability operator E, since

$$\begin{aligned} xyx &= \overrightarrow{T}(E(x,x)|T(E(x,y),E(x,y)) \\ &= T(E(x,y),E(x,y)), \end{aligned}$$

the T indistinguishability operator E can be recovered from the betweenness relation: $E(x,y) = xyx$.

Moreover, the following Proposition 6.31 shows that every fuzzy betweenness relation on a set X determines a T-indistinguishability operator E on X.

Lemma 6.30. *For every three elements x, y, z of X the following inequality holds.*

$$xyx \leq xyz.$$

Proof. From 6.25.3.a) it follows that

$$xyx = T(xyx, xxz) \leq xyz.$$

Proposition 6.31. *Let T be a continuous strict Archimedean t-norm and $f : [0,1] \to [0,1]$ the function defined by*

$$f(x) = y \text{ if and only if } x = T(y,y) \text{ i.e. } y = x_T^{\frac{1}{2}}.$$

If xyz is a fuzzy betweenness relation on a set X, then the fuzzy relations E' and E on X defined by $E'(x,y) = xyx$ and $E(x,y) = f(xyx)$ are T-indistinguishability operators on X.

Proof. It is easy to prove that if E' is a T-indistinguishability operator separating points, then E is also such an operator. Therefore it is only necessary to prove the proposition for E'.

- Reflexivity: $E'(x,x) = xxx = 1$ by 6.25.1.
- Symmetry:

$$\begin{aligned} E'(x,y) = xyx &= T(xyx, xxy) \\ &\leq yxy = E'(y,z) \end{aligned}$$

  by 6.25.3.b).
- Transitivity:

$$\begin{aligned} T(E'(x,y), E'(y,z)) &= T(xyx, yzy) \\ &\leq T(yxz, yzy) \\ &\leq T(yxz, yzx) \\ &\leq xzx = E'(x,z) \end{aligned}$$

  by Lemma 6.30.

- Separability: If $x \neq y$, then $E'(x,y) = xyx < 1$.

Remark. The initial fuzzy betweenness relation can be recovered via E as in Proposition 6.29 if and only if $E(x,y) \neq 0 \ \forall x, y \in X$.

If E is a T-indistinguishability operator on a finite set X of cardinality X with high dimension but generating a fuzzy betweenness relation on X with many values close to 1, then there will be a T-indistinguishability operator E' on X close to E defining a crisp betweenness relation on X with high cardinality, this meaning that the dimension of E' is small. In this case, instead of storing E it will be preferable to store a basis of E'.

Example 6.32. Let T be the product t-norm and E the T-indistinguishability operator on the set $X = \{1,2,3,4,5\}$ of cardinality 5 given by the following matrix

$$\begin{pmatrix} 1 & 0.74 & 0.67 & 0.50 & 0.41 \\ 0.74 & 1 & 0.87 & 0.65 & 0.53 \\ 0.67 & 0.87 & 1 & 0.74 & 0.60 \\ 0.50 & 0.65 & 0.74 & 1 & 0.80 \\ 0.41 & 0.53 & 0.60 & 0.80 & 1 \end{pmatrix}.$$

The associated fuzzy betweenness relation is given by the table 6.1.

Table 6.1 Fuzzy betweenness relation generated by E.

x	y	z	xyz	x	y	z	xyz	x	y	z	xyz	x	y	z	xyz
1	2	3	0.96	2	3	1	0.79	3	4	1	0.55	4	5	1	0.66
1	2	4	0.96	2	3	4	0.99	3	4	2	0.55	4	5	2	0.65
1	2	5	0.96	2	3	5	0.98	3	4	5	0.99	4	5	3	0.65
1	3	2	0.79	2	4	1	0.44	3	5	1	0.37	5	1	2	0.57
1	3	4	0.99	2	4	3	0.55	3	5	2	0.37	5	1	3	0.46
1	3	5	0.98	2	4	5	0.98	3	5	4	0.65	5	1	4	0.26
1	4	2	0.44	2	5	1	0.29	4	1	2	0.57	5	2	1	0.96
1	4	3	0.55	2	5	3	0.37	4	1	3	0.45	5	2	3	0.77
1	4	5	0.98	2	5	4	0.65	4	1	5	0.26	5	2	4	0.46
1	5	2	0.29	3	1	2	0.57	4	2	1	0.96	5	3	1	0.98
1	5	3	0.37	3	1	4	0.45	4	2	3	0.76	5	3	2	0.98
1	5	4	0.66	3	1	5	0.46	4	2	5	0.43	5	3	4	0.56
2	1	3	0.57	3	2	1	0.96	4	3	1	0.99	5	4	1	0.98
2	1	4	0.57	3	2	4	0.76	4	3	2	0.99	5	4	2	0.98
2	1	5	0.57	3	2	5	0.77	4	3	5	0.56	5	4	3	0.99

The dimension of E is 3. Nevertheless it is close to the one-dimensional T-indistinguishability operator E' with matrix

$$\begin{pmatrix} 1 & 0.76 & 0.67 & 0.50 & 0.40 \\ 0.76 & 1 & 0.88 & 0.68 & 0.53 \\ 0.67 & 0.88 & 1 & 0.75 & 0.60 \\ 0.50 & 0.66 & 0.75 & 1 & 0.80 \\ 0.40 & 0.53 & 0.60 & 0.80 & 1 \end{pmatrix}$$

whose associated fuzzy betweenness relation is shown in table 6.2.

Table 6.2 Fuzzy betweenness relation generated by E'.

x	y	z	xyz	x	y	z	xyz	x	y	z	xyz	x	y	z	xyz
1	2	3	1	2	3	1	0.78	3	4	1	0.56	4	5	1	0.64
1	2	4	1	2	3	4	1	3	4	2	0.56	4	5	2	0.64
1	2	5	1	2	3	5	1	3	4	5	1	4	5	3	0.64
1	3	2	0.78	2	4	1	0.43	3	5	1	0.36	5	1	2	0.58
1	3	4	1	2	4	3	0.56	3	5	2	0.36	5	1	3	0.45
1	3	5	1	2	4	5	1	3	5	4	0.64	5	1	4	0.25
1	4	2	0.43	2	5	1	0.28	4	1	2	0.58	5	2	1	1
1	4	3	0.56	2	5	3	0.36	4	1	3	0.45	5	2	3	0.78
1	4	5	1	2	5	4	0.64	4	1	5	0.25	5	2	4	0.43
1	5	2	0.28	3	1	2	0.58	4	2	1	1	5	3	1	1
1	5	3	0.36	3	1	4	0.45	4	2	3	0.78	5	3	2	1
1	5	4	0.64	3	1	5	0.45	4	2	5	0.43	5	3	4	0.56
2	1	3	0.58	3	2	1	1	4	3	1	1	5	4	1	1
2	1	4	0.58	3	2	4	0.78	4	3	2	1	5	4	2	1
2	1	5	0.58	3	2	5	0.78	4	3	5	0.56	5	4	3	1

The crisp part of the fuzzy betweenness relation generated by E' is a linear betweenness relation, since its cardinality is $2 \cdot \binom{5}{3}$.

7

Dimension and Basis

The Representation Theorem 2.54 states that every T-indistinguishability operator on a universe X can be generated by a family of fuzzy subsets of X. Nevertheless, there is no uniqueness in the selection of the family. Different families, even having different cardinalities, can generate the same operator. This point lends great interest to the theorem, since if we interpret the elements of the family as degrees of matching between the elements of the universe X and a set of prototypes, we can choose different features in order to establish this matching, thereby giving different semantic interpretations to the same T-indistinguishability operator.

Among the generating families of a T-indistinguishability operator E, the ones with low cardinality are of special interest, since they have an easy semantic interpretation and also because the information contained in the matrix representing E can be packed into a few (and sometimes just one) fuzzy subsets. Low-dimensional T-indistinguishability operators are especially desirable since all of their information can be stored in a few fuzzy subsets. These fuzzy subsets can be interpreted as the degrees of satisfiability of some features or the degree of similarity to some prototypes. A small number of fuzzy subsets will provide a better understanding of E.

This chapter studies minimal generating families (basis) and solves the problem of finding them for the minimum and for continuous Archimedean t-norms in finite universes. The case of the minimum is solved combinatorially, whereas in the Archimedean case, a geometric interpretation of the set of generators or extensional fuzzy subsets is exploded.

A reflexive and symmetric fuzzy relation can be an indistinguishability operator for different t-norms. For example, if it is a T-indistinguishability operator for a t-norm T, then it is also a T'-indistinguishability operator for any t-norm $T' \geq T$. In Section 7.3 wed determine when a reflexive and symmetric fuzzy relation is a one dimensional T-indistinguishability operator for some continuous Archimedean t-norms.

J. Recasens: Indistinguishability Operator, STUDFUZZ 260, pp. 125–145.
springerlink.com

This chapter concludes with a method based on Saaty's reciprocal matrices to find a one dimensional T-indistinguishability operator close to a given one for T the Product t-norm.

7.1 Dimension and Basis: The Archimedean Case

We deal with the problem of searching for a basis of a given T-indistinguishability operator E with respect to a continuous Archimedean t-norm T. The case when the t-norm is the minimum will be studied in the next section. Let us recall that a basis of E is a generating family of E with its set if indexes of the smallest cardinality (which is called the dimension of E).

A natural geometric representation of the set of generators H_E of a T-indistinguishability operator E defined on a set X will be exploded in order to calculate explicitly a basis of E, provided that T is an Archimedean t-norm and X finite.

H_E will be identified with a geometric subset of $[0,1]^X$. When X is finite of cardinality n and the t-norm is the product or the t-norm of Łukasiewicz, H_E is a very simple polyhedron. We will see that a basis can be chosen with all its elements in the edges of this polyhedron.

Firstly we will define a (crisp) partial ordering on H_E. The maximal elements of this partial ordering will play a special role, since the elements of the edges are maximal.

Definition 7.1. *Let E be a T-indistinguishability on a set X. In H_E we define the following relation $\leq_H$*

$$\mu \leq_H \nu \text{ if and only if } E_\mu \geq E_\nu.$$

where as always E_μ is the T-indistinguishability operator generated by μ ($E_\mu(x,y) = E_T(\mu(x), \mu(y))$).

Lemma 7.2. *$\leq_H$ is a reflexive and transitive relation.*

Proof. Trivial.

We define an equivalence relation $\sim$ on H_E in order to obtain a partial ordering:

$$\mu \sim \nu \text{ if and only if } E_\mu = E_\nu.$$

Definition 7.3. *The quotient set $H_E/\sim$ will be denoted H_E^p and $\mu \in H_E$ will be called maximal if and only if its equivalence class $\overline{\mu}$ is maximal on H_E^p.*

The next results show the relevance of maximal elements of H_E.

Lemma 7.4. *Let M be the set of maximal elements of H_E. Then M is a generating family of E.*

Proof. Trivial.

Proposition 7.5. *Let $(\mu_i)_{i\in I}$ be a generating family of E. Then there exists a generating family $(\mu_i')_{i\in I}$ of maximal generators with the same index set.*

Proof. Every μ_i is contained into a maximal one μ_i'.

Corollary 7.6. *It is always possible to find a basis of maximal elements for a given T-indistinguishability operator.*

Let us order the elements of the finite set X (i.e., $X = \{a_1, a_2, ..., a_n\}$). Then, every fuzzy subset μ of X can be identified with the point $(\mu(a_1), \mu(a_2), ..., \mu(a_n))$ of $[0,1]^n$.

If μ is a generator of a given T-indistinguishability operator E on X, then $E_\mu \geq E$ or, in a more explicit way,

$$E_T(\mu(a_i), \mu(a_j))) \geq E(a_i, a_j) \ \forall i, j = 1, 2.....n.$$

Proposition 7.7. *The set of generators of a T-indistinguishability operator E on X is the solution of the following system of inequalities:*

$$\overrightarrow{T}(\max(x_i, x_j)| \min(x_i, x_j)) \leq E(a_i, a_j) \quad 0 \leq x_i, x_j \leq 1 \quad i, j = 1, 2, ..., n.$$

This system becomes especially simple for the Product, the minimum and the Łukasiewicz t-norms.

Proposition 7.8. *If T is the Product t-norm, then H_E is the polyhedron solution of the system of inequalities*

$$x_i - E(a_i, a_j) \cdot x_j \leq 0 \quad 0 \leq x_i, x_j \leq 1 \quad i, j = 1, 2, ..., n.$$

Proposition 7.9. *If T is the Łukasiewicz t-norm, then H_E is the polyhedron solution of the system of inequalities*

$$x_i - x_j \leq 1 - E(a_i, a_j) \quad 0 \leq x_i, x_j \leq 1 \quad i, j = 1, 2, ..., n.$$

Proposition 7.10. *If T is the minimum t-norm, then H_E is the solution of system of inequalities*

$$\min(x_i, x_j) \geq E(a_i, a_j) \quad x_i \neq x_j \quad 0 \leq x_i, x_j \leq 1 \quad i, j = 1, 2, ..., n.$$

For the Product and the Łukasiewicz t-norms the location of the elements of a basis is relatively strict as it is shown in the following proposition.

Proposition 7.11. *If T is the Product or the Łukasiewicz t-norm, then the elements of a basis of a T-indistinguishability operator E are located in the (hyper) faces of H_E.*

Proof. Let $H = \{\mu_1, \mu_s, ..., \mu_k\}$ be a basis of E. If, for instance, μ_i does not lie on any face of H_E, then

$$E_\mu(x_i, x_j) > E(x_i, x_j) \; 1 \leq i < j \leq n.$$

Therefore, the set $H - \{\mu_i\}$ will also be a basis of E, which contradicts the minimality of the cardinality of H.

Proposition 7.12. *If T is the product or the Łukasiewicz t-norm, then it is always possible to find a basis of E with all its elements on the edges of H_E.*

Proof. Let $H = \{\mu_1, \mu_2, ..., \mu_k\}$ be a basis of E and let us suppose that μ_1 belongs, for example, to t faces. The set $H' = \{\mu_1', \mu_2, ..., \mu_k\}$ obtained from H by replacing μ_1' by a fuzzy set situated in an edge and contained in these t faces is also a basis of E. This replacement can be done for all $\mu_i \in H$.

It is clear that the elements of a preceding basis belong to different edges and, since the number of edges is finite, from Proposition 7.12 a method to calculate a basis of E can be derived.

Procedure to calculate a basis of a T-indistinguishability E on a finite set X(cardinality of $X = n$) for T the Łukasiewicz or the Product t-norm:

1. Calculate the edges of the set H_E.
2. $Count = 1$.
3. Build a set A obtained taking a generator from each edge of H_E.
4. Define $B(Count)$ = the set of subsets of A of $Count$ elements.
5. Select a set H of $B(Count)$ and build the T-indistinguishability operator E_H generated by H.
6. If $E_H = E$ then end.
7. Do step 5 and step 6 for all different elements of $B(Count)$.
8. $Count = Count + 1$. Go to 4.

Example 7.13. Let us consider the Product-indistinguishability operator E on a set X of cardinality 4 represented by the matrix

$$\begin{pmatrix} 1 & 0.12 & 0.41 & 0.13 \\ 0.12 & 1 & 0.12 & 0.23 \\ 0.41 & 0.12 & 1 & 0.27 \\ 0.13 & 0.23 & 0.27 & 1 \end{pmatrix}$$

It can be shown that the cardinality of the set of the edges is bounded by the number $\binom{2(n-1)}{n-1}$ being n the cardinality of X. The set of edges in this case is

$$\begin{array}{ll}
\{(0.12\ 1.00\ 0.29\ 0.23), & (0.41\ 0.12\ 1.00\ 0.52), \\
(0.12\ 1.00\ 0.25\ 0.92), & (0.41\ 0.12\ 1.00\ 0.27), \\
(0.12\ 1.00\ 0.29\ 0.92), & (0.66\ 0.23\ 0.27\ 1.00), \\
(0.12\ 1.00\ 0.12\ 0.44), & (1.00\ 0.12\ 0.48\ 0.13), \\
(0.12\ 1.00\ 0.12\ 0.23), & (1.00\ 0.12\ 0.41\ 0.13), \\
(0.13\ 0.23\ 0.32\ 1.00), & (1.00\ 0.12\ 0.41\ 0.52), \\
(0.13\ 0.23\ 0.27\ 1.00), & (1.00\ 0.12\ 1.00\ 0.27), \\
(0.29\ 1.00\ 0.12\ 0.44), & (1.00\ 0.12\ 1.00\ 0.52), \\
(0.29\ 1.00\ 0.12\ 0.23), & (1.00\ 0.57\ 0.41\ 0.13), \\
(0.35\ 1.00\ 0.85\ 0.23), & (1.00\ 0.57\ 0.48\ 0.13)\}
\end{array}$$

and a basis is

$$\{(0.12\ 1.00\ 0.25\ 0.92),\ (0.29\ 1.00\ 0.12\ 0.23)\}.$$

When the cardinality of the universe of discourse X is 3, there is a nice geometric interpretation of these results.

Example 7.14. The set of generators H_E of the Product-indistinguishability operator E on $X = \{a_1, a_2, a_3\}$ with matrix

$$\begin{pmatrix} 1 & 0.23 & 0.37 \\ 0.23 & 1 & 0.26 \\ 0.37 & 0.26 & 1 \end{pmatrix}$$

is the part of the pyramid with vertex on the origin of coordinates with edges passing through the points A, B, C, D, E, F contained in $[0, 1]^3$.

$$\begin{array}{ll}
A = (0.37, 0, 26, 1) & B = (1, 0.23, 0.86) \\
C = (1, 0.23, 0.37) & D = (0.72, 1, 0.26) \\
E = (0.23, 1, 0.26) & F = (0.23, 1, 0.61).
\end{array}$$

A basis of E is given by the two fuzzy subsets

$$B = (1, 0.23, 0.86)\ F = (0.23, 1, 0.61)$$

and E is bidimensional. See Figure 7.1.

Example 7.15. The set of generators H_E of the T-indistinguishability operator E on $X = \{a_1, a_2, a_3\}$ (T the Łukasiewicz t-norm) with matrix

$$\begin{pmatrix} 1 & 0.32 & 0.42 \\ 0.32 & 1 & 0.36 \\ 0.42 & 0.36 & 1 \end{pmatrix}$$

is the part of the prism with edges parallel to the line $x = y = z$ passing through the points A, B, C, D, E, F contained in $[0, 1]^3$.

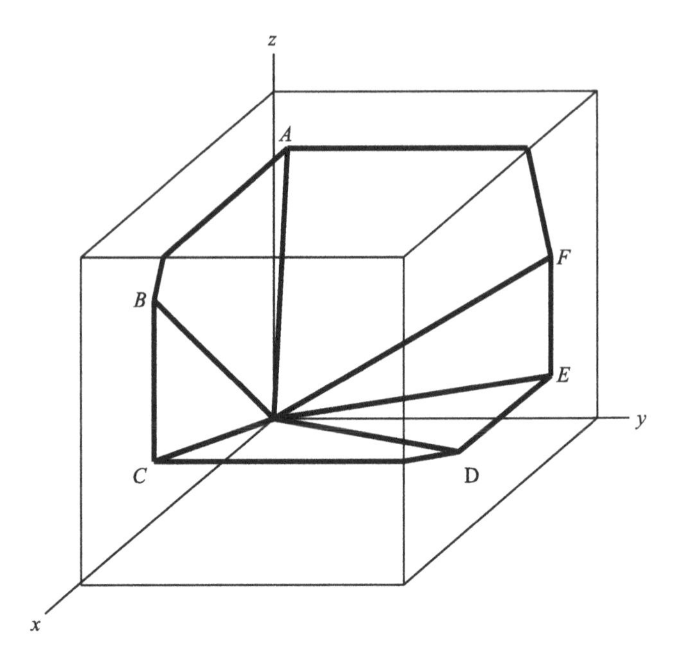

Fig. 7.1 H_E of Example 7.14

$$A = (0.68, 0, 0.64)\ B = (0.68, 0, 0.1)$$
$$C = (0.58, 0.64, 0)\ D = (0, 0.68, 0.04)$$
$$E = (0, 0.68, 0.58)\ F = (0.06, 0, 0.64).$$

A basis of E is given by the two fuzzy subsets

$$A = (0.68, 0, 0.64)\ B = (0.68, 0, 0.1)$$

and E is bidimensional. See Figure 7.2.

The results obtained for the Product and the Łukasiewicz t-norms can be extended to any continuous Archimedean t-norm applying the results obtained in Section 4.2 of Chapter 4. It has been proved there that if f is an isomorphism between two t-norms T and T' and E is a T-indistinguishability operator on a set X of dimension k having $\{\mu_1, \mu_2, ..., \mu_k\}$ as a basis, then $E' = f \circ E$ is a T'-indistinguishability operator on X of dimension k and $\{f \circ \mu_1, f \circ \mu_2, ..., f \circ \mu_k\}$ is a basis of E'.

So let T be a continuous strict Archimedean t-norm and f an isomorphism between T and the product t-norm. If E is a T-indistinguishability operator on a finite set X, then $E' = f \circ E$ is a Product-indistinguishability operator and we can calculate its dimension k an a basis $\{\mu_1, \mu_2, ..., \mu_k\}$. E has also dimension k and $\left\{f^{-1} \circ \mu_1, f^{-1} \circ \mu_2, ..., f^{-1} \circ \mu_k\right\}$ is a basis of E.

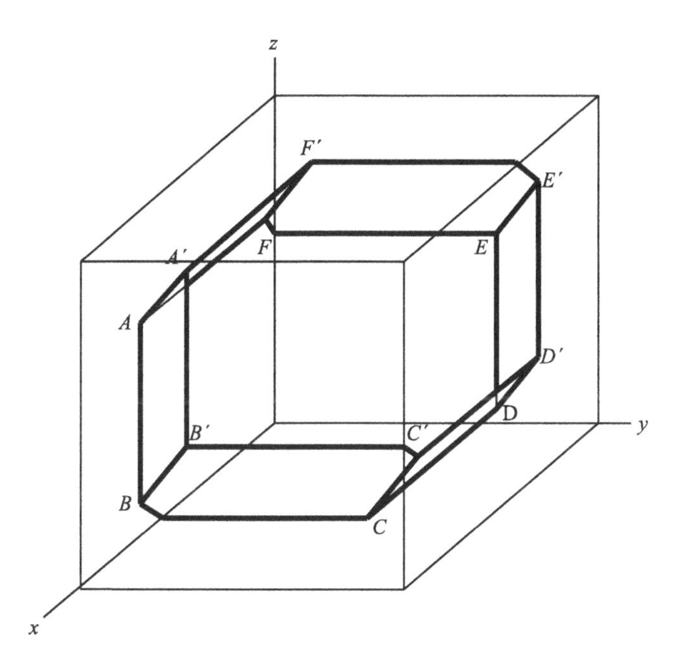

Fig. 7.2 H_E of Example 7.15

A similar result applies for non-strict continuous Archimedean t-norms, since they are isomorphic to the Łukasiewicz t-norm.

The geometric interpretation of the set H_E also gives a new theoretical method to obtain the transitive closure of a reflexive and symmetric relation when the t-norm is continuous Archimedean. It is not an efficient method and hence it is not appropriate for practical purposes. Its interest lies in the geometric interpretation of the transitive closure and in its relation with H_E.

We can restrict our study to the Product and the Łukasiewicz t-norms. Both cases are similar and we will only study the latter.

Let R be a reflexive and symmetric fuzzy relation on a finite set $X = \{a_1, a_2, ..., a_n\}$ of cardinality n and T the Łukasiewicz t-norm. We will calculate the set $H_{\overline{R}}$ of generators of its T-transitive closure $\overline{R}$.

Let us recall that $R \leq \overline{R}$ and if E is another T-indistinguishability operator on X satisfying $R \leq E$, then $\overline{R} \leq E$.

According to Proposition 7.9, $H_{\overline{R}}$ is the polyhedron solution of the system of inequalities

$$x_i - x_j \leq 1 - \overline{R}(a_i, a_j) \quad 0 \leq x_i, x_j \leq 1 \quad i, j = l, 2, ..., n.$$

But this system is equivalent to

$$x_i - x_j \leq 1 - R(a_i, a_j) \quad 0 \leq x_i, x_j \leq 1 \quad i, j = l, 2, ..., n.$$

Therefore from R we can calculate the set $H_{\overline{R}}$ and from here we can calculate $\overline{R}$ finding a basis and then generating $\overline{R}$ from it.

Note that the inequalities

$$x_i - x_j < 1 - R(a_i, a_j) \text{ with } R(a_i, a_j) < \overline{R}(a_i, a_j)$$

are superfluous and therefore the numbers $\overline{R}(a_i, a_j)$ that are greater than $R(a_i, a_j)$ are $\mathbb{Q}$-linear combination of the $\overline{R}(a_i, a_j)$ that coincide with their respective $R(a_i, a_j)$. Therefore, the more numbers $R(a_i, a_j)$ different from their corresponding $\overline{R}(a_i, a_j)$, the less edges will have $H_{\overline{R}}$ and the smaller its dimension. In other words,

> The farther a reflexive and symmetric relation R is from its transitive closure $\overline{R}$, the smaller the dimension of $\overline{R}$.

Note that this rule of thumb coincides with the one of the introduction of Chapter 6.

7.2 Dimension and Basis: The Minimum t-Norm

The calculation of the dimension and a basis in the Archimedean case has been of geometric nature. For the min-indistinguishability operators the calculation will be combinatorial.

We will assume separability for all min-indistinguishability operators in this section. Otherwise, if $\sim$ is the equivalence relation on a finite set X $x \sim y$ if and only if $E(x, y) = 1$, then the induced similarity relation $\overline{E}$ on the quotient set $\overline{X}/\sim$ separates points. Moreover, a fuzzy subset μ of X is extensional with respect to E if and only if the induced fuzzy subset $\overline{\mu}$ of $\overline{X}$ is extensional with respect to $\overline{E}$. So E and $\overline{E}$ have the same dimension and $(\mu_i)_{i \in I}$ is a basis of E if and only if $(\overline{\mu}_i)_{i \in I}$ is a basis of $\overline{E}$.

Proposition 7.16. *Let E be a* min*-indistinguishability operator on a finite set X. E is one dimensional if and only if E has a column μ_x which is a one to one map.*

Proof

$\Rightarrow$) Let μ be a basis of E. μ is a one to one map since E separates points. Let $a \in X$ satisfy $\mu(a) = \max_{x \in X}\{\mu(x)\}$. The fuzzy subset μ' of X defined by

$$\mu'(x) = \begin{cases} \mu(x) & \text{if } x \neq a \\ 1 & \text{if } x = a \end{cases}$$

is a column and a basis of E.

$\Leftarrow$) Let μ_x be a column of E which is one to one. Let us prove that $E_{\mu_x} = E$. If $y \neq z$, then

$$\begin{aligned} E_{\mu_x}(y, z) &= E_{\min}(\mu_x(y), \mu_x(z)) = \min(\mu_x(y), \mu_x(z)) \\ &= \min(E(x, y), E(x, z)) \leq E(y, z). \end{aligned}$$

But since μ_x is extensional with respect to E, $E_{\mu_x} \geq E$ and equality holds.

The following definitions will be useful for calculating the dimension and a basis of a min-indistinguishability operator.

Definition 7.17. *A column μ_x of a* min*-indistinguishability operator E on a finite set X is maximal if and only if for any other column μ_y of E*

$$|\{\mu_x(z), z \in X\}| \geq |\{\mu_y(z), z \in X\}|$$

where $|A|$ denotes the cardinality of the set A.

So a maximal column has a maximal number of different values. H_M will denote the set of maximal columns of E.

If μ_x is a maximal column and $\alpha \in \{\mu_x(y), y \in X\}$, then X^α is the subset of X defined by $X^\alpha = \{y \in X$ such that $\mu_x(y) = \alpha\}$ and $|X^\alpha|$ will be called the frequency ($\text{fr}(\alpha)$) of α in μ_x.

Definition 7.18. *The order $\theta(\mu_x)$ of a maximal column μ_x is the maximum of the frequencies of the values in μ_x, i.e.*

$$\theta(\mu_x) = \max\{\text{fr}(\mu_x(y)),\ y \in X\}.$$

Definition 7.19. *The order $\theta(E)$ of a* min*-indistinguishability operator E on a set X is the minimum of the orders of its maximal columns.*

$$\theta(E) = \min\{\theta(\mu_x),\ x \in X\}.$$

Example 7.20. Let E be the min-indistinguishability operator on a set X of cardinality 10 with the matrix 7.1. The maximal columns are printed in boldface and their orders are all 7. Therefore the order of E is $\theta(E) = 7$.

Table 7.1 Matrix E

	x_1	x_2	x_3	x_4	x_5	x_6	x_7	x_8	x_9	x_{10}
x_1	**1**	0.10	**0.24**	0.10	**0.10**	**0.10**	0.10	0.10	0.12	0.10
x_2	**0.10**	1	**0.10**	0.10	**0.10**	**0.10**	0.10	0.10	0.10	0.10
x_3	**0.24**	0.10	**1**	0.10	**0.10**	**0.10**	0.10	0.10	0.12	0.10
x_4	**0.10**	0.10	**0.10**	1	**0.10**	**0.10**	0.33	0.10	0.10	0.10
x_5	**0.10**	0.10	**0.10**	0.10	**1**	**0.42**	0.10	0.10	0.10	0.17
x_6	**0.10**	0.10	**0.10**	0.10	**0.42**	**1**	0.10	0.10	0.10	0.17
x_7	**0.10**	0.10	**0.10**	0.33	**0.10**	**0.10**	1	0.10	0.10	0.10
x_8	**0.10**	0.10	**0.10**	0.10	**0.10**	**0.10**	0.10	1	0.10	0.10
x_9	**0.12**	0.10	**0.12**	0.10	**0.10**	**0.10**	0.10	0.10	1	0.10
x_{10}	**0.10**	0.10	**0.10**	0.10	**0.17**	**0.17**	0.10	0.10	0.10	1

Lemma 7.21. *If $(\mu_i)_{i\in I}$ is a generating family of a* min*-indistinguishability operator on a set X, then there exists another family of generators $(\mu'_i)_{i\in I}$ with the same cardinality such that for any $i \in I$ and $x \in X$ $\mu'_i(x) \in \mathrm{Im}(E)$.*

The next proposition gives an upper bound to the dimension of a min-indistinguishability operator E and allows us to build an algorithm for the search of a basis of E.

Proposition 7.22. [63] *If $\theta(E)$ is the order of a* min*-indistinguishability operator on a set X and σ is the smallest natural number satisfying $2^\sigma \geq \theta(E)$, then* $\dim(E) \leq \sigma + 1$.

From this proposition two algorithms will be developed. The second one is simpler than the first one but gives a generating family of cardinality at most $\dim(E) + 1$, while the first one gives always a basis of E.

The first **algorithm** goes as follows

1. Compute the orders of the columns of E and select the maximal ones.
2. Compute the order $\theta(E)$ of E and the upper bound of the dimension $(\sigma + 1)$.
3. Select a maximal column μ_x whose order is $\theta(E)$.
4. Build $\sigma + 1$ column vectors μ_i of dimension n and initialize them with 1's.
5. For any unrepeated value $\mu_x(y)$ in μ_x do $\mu_i(y) \leftarrow \mu_x(y)$ for i from 1 to $\sigma + 1$. Do this assignment for all unrepeated values in μ_x.
6. Select a repeated value α in μ_x and find its associate X^α.
7. For any pair x_i, x_j of elements of X^α such that $E(x_i, x_j) = \alpha' \neq \alpha$ do $\mu_1(x_i) = \mu_1(x_j) \leftarrow \alpha$, $\mu_2(x_i \leftarrow 1$, $\mu_2(x_j) \leftarrow \alpha'$.
8. For any element $x_r \in X^\alpha$ not selected in the preceding step do $\mu_i(x_r) \leftarrow \alpha$ for $i = 1, 2, ...,$ until for any pair of elements x_r, x_s included in this step there exists an i such that $\mu_i(x_r) \neq \mu_i(x_s)$.
9. If there exists in μ_x another repeated value, select it and go to step 6.
10. For any pair of elements x', x'' with different repeated values α, α' such that $\mu_i(x') = \mu_i(x'')$, select j such that $\mu_j(x') = \mu_j(x'') = 1$ and do $\mu_j(x') \leftarrow E(x', x'')$.
11. Delete all constant columns.
12. End.

In the next example, a basis for a given min-indistinguishability operator E is found using the previous algorithm.

Example 7.23. Let E be the min-indistinguishability operator defined in Example 7.20.

1. The maximal columns are the 3rd and 8th ones (printed in boldface), both of order 2.
2. $\theta(E) = 2$. The upper bound of the dimension is 2.

3. We select, for example, the 8th column.
4. Build two initialized vectors.
5. List the unrepeated values like in table 7.2.
6. Select the repeated value $\alpha_1 = 0.10$, $X^{\alpha_1} = \{x_1, x_{10}\}$. Do the same with the value $\alpha_2 = 0.13$, $X^{\alpha_2} = \{x_7, x_9\}$.
7. 8. 9. 10. Since $E(x_1, x_{10}) = 0.25 \neq 0.10$, place them in the second column and do the same with the value $\alpha_2 = 0.13$, because $E(x_7, x_9) = 0.28 \neq 0.13$ See table 7.3.

Table 7.2

	x_1	x_2	x_3	x_4	x_5	x_6	x_7	x_8	x_9	x_{10}
μ_1	1	0.27	0.40	0.35	0.36	0.30	1	1	1	1
μ_2	1	0.27	0.40	0.35	0.36	0.30	1	1	1	1

Table 7.3

	x_1	x_2	x_3	x_4	x_5	x_6	x_7	x_8	x_9	x_{10}
μ_1	0.10	0.27	0.40	0.35	0.36	0.30	0.13	1	0.13	1
μ_2	0.25	0.27	0.40	0.35	0.36	0.30	0.28	1	1	1

The second **algorithm** following Proposition 7.22 is:

1. Counter $\leftarrow 0$, $|X| = n \leftarrow N$.
2. Compute the columns' orders and select the maximal ones.
3. Compute the order $\theta(E)$ of E and calculate σ.
4. Select a maximal column μ_x of E whose order equals the order of E.
5. Counter $\leftarrow$ Counter+1 ($C = C + 1$).
6. If $C = 1$, then build $\sigma + 1$ columns μ_i of dimension N, initialize them with 1's and do $\mu_1(y) = \mu_x(y)$ for any $y \in X$. Go to 7 otherwise.
7. Select a repeated value α of μ_x and find the set X^α associated to α.
8. If $\sigma = 1$ End.
9. Select the min-indistinguishability operator $E' = E|X^\alpha$, do $E \leftarrow E'$, $X \leftarrow X^\alpha$, $N \leftarrow |X^\alpha|$.
10. If $\theta(E) = n - 1$, then go to 13.
11. Do $\mu_C(y) \leftarrow \mu_x(y)$ for any $y \in X$ such that $\mu_x(y) \neq \alpha$.
12. For any element x not being selected in 11 do $\mu_j(y) \leftarrow \alpha$ for $j = 1, 2, ...$ until for any pair of elements x_r, x_s selected in this step there exists a j with $\mu_j(x_r) \neq \mu_j(x_s)$ and $C \leq j \leq \sigma + 1$.
13. Select another repeated value and go to 5.
14. End.

Example 7.24. In this example the algorithm calculates a basis for the min-indistinguishability operator defined on a set X of cardinality 10 given in Table 7.4.

Table 7.4 Matrix E

	x_1	x_2	x_3	x_4	x_5	x_6	x_7	x_8	x_9	x_{10}
x_1	1	0.10	0.12	0.12	0.11	0.12	0.12	0.10	0.12	0.12
x_2	0.10	1	0.10	0.10	0.10	0.10	0.10	0.22	0.10	0.10
x_3	0.12	0.10	1	0.36	0.11	0.13	0.23	0.10	0.16	0.19
x_4	0.12	0.10	0.36	1	0.11	0.13	0.23	0.10	0.16	0.19
x_5	0.11	0.10	0.11	0.11	1	0.11	0.11	0.10	0.11	0.11
x_6	0.12	0.10	0.13	0.13	0.11	1	0.13	0.10	0.13	0.13
x_7	0.12	0.10	0.23	0.23	0.11	0.13	1	0.10	0.16	0.19
x_8	0.10	0.22	0.10	0.10	0.10	0.10	0.10	1	0.10	0.10
x_9	0.12	0.10	0.16	0.16	0.11	0.13	0.16	0.10	1	0.16
x_{10}	0.12	0.10	0.19	0.19	0.11	0.13	0.19	0.10	0.16	1

The maximal columns are the 3rd and the 4th,both of order 2. $\theta(E) = 2$ and the upper bound to the dimension is also 2.

In this case the algorithm selects the basis of E shown in Table 7.5.

Table 7.5

	x_1	x_2	x_3	x_4	x_5	x_6	x_7	x_8	x_9	x_{10}
μ_1	0.12	0.10	0.36	1	0.11	0.13	0.23	0.10	0.16	0.19
μ_2	0.12	0.22	0.36	1	0.11	0.13	0.23	1	0.16	0.19

Example 7.25. In this example the algorithm produces a set of generators of cardinality $\dim(E) + 1$. E is given in Table 7.6.

Table 7.6 Matrix E

	x_1	x_2	x_3	x_4	x_5	x_6	x_7	x_8	x_9	x_{10}
x_1	1	0.10	0.10	0.10	0.10	0.10	0.10	0.10	0.10	0.10
x_2	0.10	1	0.10	0.10	0.10	0.18	0.10	0.10	0.10	0.10
x_3	0.10	0.10	1	0.17	0.17	0.10	0.17	0.23	0.17	0.23
x_4	0.10	0.10	0.17	1	0.17	0.10	0.17	0.17	0.10	0.17
x_5	0.10	0.10	0.17	0.17	1	0.10	0.17	0.17	0.10	0.17
x_6	0.10	0.18	0.10	0.10	0.10	1	0.10	0.10	0.10	0.10
x_7	0.10	0.10	0.17	0.17	0.17	0.10	1	0.17	0.10	0.17
x_8	0.10	0.10	0.23	0.17	0.17	0.10	0.17	1	0.10	0.25
x_9	0.10	0.10	0.10	0.10	0.10	0.10	0.10	0.10	1	0.10
x_{10}	0.10	0.10	0.23	0.17	0.17	0.10	0.17	0.25	0.10	1

The algorithm gives a set of three generators (Table 7.7) while Table 7.8 gives a basis of E.

Table 7.7 A generating family of E with three fuzzy subsets

	x_1	x_2	x_3	x_4	x_5	x_6	x_7	x_8	x_9	x_{10}
μ_1	0.10	0.10	0.23	0.17	0.17	0.10	0.17	0.25	0.10	1
μ_2	0.10	0.18	0.23	0.17	0.17	1	1	0.25	0.10	1
μ_3	0.10	0.18	0.23	0.17	1	1	1	0.25	1	1

Table 7.8 A basis of E

	x_1	x_2	x_3	x_4	x_5	x_6	x_7	x_8	x_9	x_{10}
ν_1	1	0.10	0.23	1	0.17	0.10	0.17	0.25	0.10	1
ν_2	0.10	0.18	0.23	0.17	1	1	0.17	0.25	0.10	1

7.3 One Dimensional Fuzzy Relations

A fuzzy relation E can be an indistinguishability operator for many t-norms. For instance, if E is such an operator for a given t-norm T, then it trivially is also a T'-indistinguishability operator for any t-norm T' with $T' \leq T$. The dimension of E for different t-norms may vary and hence the number of fuzzy subsets needed to generate the same operator changes for different t-norms.

The simplest case is when E can be generated by a single fuzzy set. In this section an algorithm that allows us to decide if a given fuzzy relation is a one dimensional T-indistinguishability operator for some Archimedean t-norm and to find all t-norms with this property is provided. The procedure is based on previous results in this chapter and in the betweenness relations generated by T-indistinguishability operators when T is a continuous Archimedean t-norm.

Lemma 7.26. *Let T be a continuous Archimedean t-norm and E a T-indistinguishability operator on a set X. E separates points if and only if for all $x, y \in X$, $x \neq y$ the columns μ_x and μ_y of E associated to them are also different.*

Proof.
$\Rightarrow$) $1 = E(x,x) \neq E(y,x)$.
$\Leftarrow$) Let $z \in X$ be such that $E(x,z) \neq E(y,z)$.
Since

$$T(E(x,y), E(y,z)) \leq E(x,z)$$

and

$$T(E(x,y), E(x,z)) \leq E(y,z),$$

$E(x,y) = 1$ would imply the equality between $E(x,z)$ and $E(y,z)$.

Lemma 7.27. *Let E be a one dimensional T-indistinguishability operator on set X, $\{\mu\}$ a basis of E and $x, y \in X$. $E(x,y) = 1$ if and only if $\mu(x) = \mu(y)$.*

Proof. Trivial.

As it has been seen before, if E is a T-indistinguishability operator on a set X, then the (crisp) relation on X defined by $x \sim y$ if and only if $E(x,y) = 1$ is an equivalence relation and the fuzzy relation $\overline{E}$ defined on $X/\sim$ by $\overline{E}(\overline{x},\overline{y}) = E(x,y)$ is a T-indistinguishability operator that separates points.

Proposition 7.28. *If E is one-dimensional generated by μ, then $\overline{E}$ is also one-dimensional and generated by $\overline{\mu}$ (where $\overline{\mu}$ is defined by $\overline{\mu}(\overline{x}) = \mu(x)$).*

Proof. If $E = E_\mu$, then

$$\begin{aligned}\overline{E}(\overline{x},\overline{y}) &= E(x,y) = E_\mu(x,y)\\ &= E_T(\mu(x),\mu(y))\\ &= E_T(\overline{\mu}(\overline{x}),\overline{\mu}(\overline{y})) = E_{\overline{\mu}}(\overline{x},\overline{y}).\end{aligned}$$

Due to this proposition we can restrict our study to operators that separate points.

We will need the following Proposition 7.30 that states that for finite sets linearity of a betweenness relation can be deduced from only some particular elements of the betweenness relation.

Lemma 7.29. *Let T be a continuous Archimedean t-norm, $\{\mu\}$ a basis of a one dimensional T-indistinguishability operator E on a set X separating points with $E(x,y) \neq 0$ for all $x,y \in X$ and B the linear betweenness relation generated on X by E. $(x,y,z) \in B$ if and only if $\mu(x) < \mu(y) < \mu(z)$ or $\mu(z) < \mu(y) < \mu(x)$.*

Proof. It is a consequence of Proposition 6.3.

Proposition 7.30. *Let $X = \{x_1, x_2, ..., x_n\}$ be a finite set of cardinality n and B a betweenness relation on X. If $(x_1, x_i, x_{i+1}) \in B$ for all $i = 1, 2, ..., n-1$, then B is the linear betweenness relation*

$$\begin{aligned}B = \{&(x_i, x_j, x_k) \in X \times X \times X \text{ such that } 1 \leq i < j < k \leq n\\ &\text{or } 1 \leq k < j < i \leq n\}.\end{aligned}$$

Proof. Thanks to 6.1.2 it is only necessary to prove that $(x_i, x_j, x_k) \in B$ when $1 \leq i < j < k \leq n$. It will be done by induction on n.

If $n \leq 3$, then the result is trivial.

Let us assume that the result is true for all sets of cardinality $n-1$.

If B' is the restriction of B to $X - \{x_n\}$, then B' is a linear betweenness relation and therefore $(x_i, x_j, x_k) \in B$ when $1 \leq i < j < k \leq n-1$. It only remains to prove that $(x_i, x_j, x_n) \in B$ if $1 \leq i < j < n-1$.

Case a: $i = 1$.

If $j = n-1$, then $(x_1, x_{n-1}, x_n) \in B$ since it is assumed in the hypothesis.

If $j \neq n-1$, then $(x_1, x_j, x_{n-1}) \in B' \subseteq B$ and $(x_1, x_{n-1}, x_n) \in B$.

From 6.1.4 it follows that $(x_1, x_j, x_n) \in B$.

Case b: $i \neq 1$.

$(x_1, x_i, x_j) \in B' \subseteq B$ and $(x_1, x_j, x_n) \in B$ as it is proved in Case a. From 6.1.4 it follows that $(x_i, x_j, x_n) \in B$.

Proposition 7.31. *Let T be a continuous Archimedean t-norm, E a one-dimensional T-indistinguishability operator separating points on a finite set X with $E(x,y) \neq 0$ $\forall x, y \in X$ and a, b a pair of elements of X such that $E(a,b) = \min\{E(x,y), x, y \in X\}$. Then*

1. *The pair (a,b) is unique.*
2. *The columns of a and b are basis of E.*

Proof

1. Since E is one dimensional it is generated by a fuzzy subset μ. Let $a, b \in X$ be such that $\mu(a) = \min\{\mu(x) \mid x \in X\}$ and $\mu(b) = \max\{\mu(x) \mid x \in X\}$. a and b are unique because μ is a one to one map.

   Let $x, y \in X$ with $x \neq y$ and $x, y \neq a, b$. Without loss of generality we can assume that $\mu(x) > \mu(y)$.

$$\begin{aligned} E(a,b) &= E_T(\mu(a), \mu(b)) = \overrightarrow{T}(\mu(b)|\mu(a)) \\ &< \overrightarrow{T}(\mu(x)|\mu(y)) = E(x,y). \end{aligned}$$

2. a. μ_a generates E: Let μ be a basis of E and $x \in X$.

$$\begin{aligned} \mu_a(x) &= E(a,x) = E_\mu(a,x) \\ &= E_T(\mu(a), \mu(x) = \overrightarrow{T}(\mu(x)|\mu(a)). \end{aligned}$$

Let $x, y \in X$, $x \neq y$. We can assume without loss of generality that $\mu(x) > \mu(y)$.

$$\begin{aligned} E_{\mu_a}(x,y) &= E_T(\mu_a(x), \mu_a(y)) \\ &= E_T(\overrightarrow{T}(\mu(y)|\mu(a)), \overrightarrow{T}(\mu(x)|\mu(a))) \\ &= \overrightarrow{T}(\overrightarrow{T}(\mu(y)|\mu(a)), \overrightarrow{T}(\mu(x)|\mu(a))) \\ &= \overrightarrow{T}(\mu(x)|\mu(y)) = E_\mu(x,y) = E(x,y). \end{aligned}$$

b. μ_b generates E: Let μ be a basis of E and $x \in X$.

$$\begin{aligned} \mu_b(x) &= E(b,x) = E_\mu(b,x) \\ &= E_T(\mu(b), \mu(x) = \overrightarrow{T}(\mu(b)|\mu(x)). \end{aligned}$$

Let $x, y \in X$, $x \neq y$. We can assume without loss of generality that $\mu(x) > \mu(y)$.

$$\begin{aligned}E_{\mu_b}(x,y) &= E_T(\mu_b(x),\mu_b(y))\\ &= E_T(\overrightarrow{T}(\mu(b)|\mu(x)),\overrightarrow{T}(\mu(b)|\mu(y)))\\ &= \overrightarrow{T}(\overrightarrow{T}(\mu(b)|\mu(x)),\overrightarrow{T}(\mu(b)|\mu(y)))\\ &= \overrightarrow{T}(\mu(x)|\mu(y)) = E_\mu(x,y) = E(x,y).\end{aligned}$$

Proposition 7.32. *Using the previous notations, let us order the elements of* $X = \{x_1, x_2, ..., x_n\}$ *in such a way that* $x_1 = a, x_n = b$ *and* $E(x_1, x_i) < E(x_1, x_j)$ *if and only if* $i > j$. *Then the betweenness relation* B *generated by* E *in* X *is*

$$(x_i, x_j, x_k) \in B \text{ if and only if } i < j < k \text{ or } i > j > k.$$

Corollary 7.33. *With the previous notations the following equalities hold.*

$$T(E(x_i, x_j), E(x_j, x_k)) = E(x_i, x_k)\ i < j < k.$$

Note that if t is an additive generator of T and since $E(x,y) \neq 0\ \forall x, y \in X$, the preceding equalities are equivalent to

$$t(E(x_i, x_j)) + t(E(x_j, x_k)) = t(E(x_i, x_k))\ i < j < k.$$

From Proposition 7.30 all these $\binom{n}{3}$ equalities must be the consequence of the $n - 2$ equalities

$$t(E(x_1, x_i)) + t(E(x_i, x_{i+1})) = t(E(x_1, x_{i+1}))\ i = 2, 3, ..., n-1. \quad (*)$$

This means that all the entries of E are determined by $E(x_1, x_i)$ and $E(x_i, x_{i+1})$ $i = 2, 3, ..., n - 1$. Putting these ideas together, we obtain an algorithm to decide when a given reflexive and symmetric fuzzy relation on a finite set X of cardinality n with $E(x,y) \neq 0\ \forall x, y \in X$ is a one-dimensional T-indistinguishability operator for some continuous Archimedean t-norm T. The **algorithm** also finds all continuous Archimedean t-norms with this property.

1. Order the elements of X as in Proposition 7.32.
2. Define, if possible, a function

$$\begin{aligned}t : P = \{&E(x_1, x_i) \text{ such that } i \in \{2, 3, ..., n-1\}\} \cup \{E(x_i, x_{i+1})\\ &\text{such that } i \in \{2, 3, ..., n-1\}\} \to \mathbb{R}^+\end{aligned}$$

   such that $\forall m, n \in P\ m < n \Rightarrow t(m) > t(n)$ and satisfying (*).

   (If E is one-dimensional for some continuous Archimedean t-norm T, this is always possible following this order of assignment: $t(E(x_1, x_2))$, $t(E(x_2, x_3))$, $t(E(x_1, x_3))$, $t(E(x_3, x_4))$, $t(E(x_1, x_4))$, ... If no such t exists, then E is not one-dimensional for any continuous Archimedean t-norm T.)

3. Extend t to $Q = \{E(x_i, x_j)$ such that $1 \leq i < j \leq n\}$ applying Proposition 7.30 and Corollary 7.33.
4. If $\forall m, n \in B\ m < n \Rightarrow t(m) > t(n)$, then E is one dimensional with respect to any t-norm T having an additive generator interpolating the images of Q. In any other case, E is not one-dimensional for any continuous Archimedean t-norm.

Example 7.34. We will apply the preceding algorithm to the relation E on the set $X = \{a, b, c, d, e\}$ given by the following matrix:

$$\begin{array}{c} \\ a \\ b \\ c \\ d \\ e \end{array} \begin{array}{c} \begin{array}{ccccc} a & b & c & d & e \end{array} \\ \begin{pmatrix} 1 & 0.87 & 0.77 & 0.54 & 0.49 \\ 0.87 & 1 & 0.9 & 0.67 & 0.62. \\ 0.77 & 0.9 & 1 & 0.77 & 0.72 \\ 0.54 & 0.67 & 0.77 & 1 & 0.95 \\ 0.49 & 0.62 & 0.72 & 0.95 & 1 \end{pmatrix} \end{array}.$$

$\min\{E(x, y), x, y \in X\} = E(a, e) = 0.49.$
Step 1. We order the elements of X in the following way:

$$a < b < c < d < e.$$

Step 2.

$$\begin{aligned} t(E(a, b)) &= t(0, 87) = 10 \\ t(E(b, c)) &= t(0.9) = 5 \\ t(E(a, c)) &= t(0.77) = 10 + 5 = 15 \\ t(E(c, d)) &= t(0.77) = 15 \\ t(E(a, d)) &= t(0.54) = 15 + 15 = 30 \\ t(E(d, e)) &= t(0.95) = 2 \\ t(E(a, e)) &= t(0.49) = 30 + 2 = 32. \end{aligned}$$

Step 3.

$$\begin{aligned} t(E(b, d)) &= t(0.67) = t(E(b, c)) + t(E(c, d)) = 5 + 15 = 20 \\ t(E(c, e)) &= t(0, 72) = t(E(c, d)) + t(E(d, e)) = 15 + 2 = 17 \\ t(E(b, e)) &= t(0.62) = t(E(b, c) + t(E(c, e)) = 5 + 17 = 22. \end{aligned}$$

Every decreasing map $t : [0, 1] \to [0, \infty]$ that interpolates the preceding values with $t(1) = 0$ is an additive generator of a continuous Archimedean t-norm for which E is one-dimensional (see Figure 7.3).

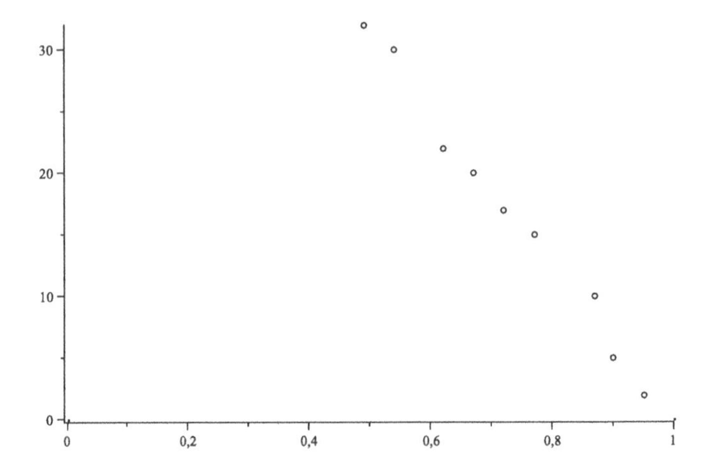

Fig. 7.3 Points to be interpolated by t in Example 7.34

It is quite obvious that every reflexive and symmetric fuzzy relation R on a finite set X with $R(x,y) \neq 1 \; \forall x, y \in X$ with $x \neq y$ is an indistinguishability operator for a suitable (maybe very small) continuous Archimedean t-norm. Nevertheless, not all such fuzzy relations can be viewed as one dimensional for every set of cardinality greater than 2.

Proposition 7.35. *A reflexive and symmetric fuzzy relation R on a set $X = \{a,b,c\}$ of cardinality 3 is a one-dimensional indistinguishability operator that separates points for some continuous Archimedean t-norm T if and only if the minimum of $R(a,b), R(b,c), R(a,c)$ is attained by only one of them.*

Proof

$\Rightarrow$) Proposition 7.31.1.

$\Leftarrow$) We can assume without loss of generality that

$$R(a,c) < R(b,c) \leq R(a,b).$$

In this case, we can assign arbitrary values

$$t(R(b,c)) > t(R(a,b)) \text{ if } R(b,c) < R(a,b)$$

or

$$t(R((b,c)) = t(R(a,b)) \text{ if } R(b,c) = R(a,b),$$

and define

$$t(R(a,c)) = t(R(a,b)) + t(R(b,c)).$$

Corollary 7.36. *On every set X of cardinality greater than 2 there are reflexive and symmetric fuzzy relations R with $R(x,y) \neq 0 \; \forall x, y \in X$ that* *are not one-dimensional indistinguishability operators for any continuous Archimedean t-norm.*

Proof. It suffices to choose a relation R on X that restricted to a subset of X of cardinality 3 does not satisfy the conditions of Proposition 7.35.

7.4 Approximating T-Indistinguishability Operators by One Dimensional Ones

In this section, using some ideas of T.L. Saaty [124] a method to generate a one dimensional T-indistinguishability operator close to a given one will be provided.

Definition 7.37. *An $n \times n$ real matrix A with entries $a_{ij} > 0$ $1 \leq i,j \leq n$ is reciprocal if and only if $a_{ij} = \frac{1}{a_{ji}}$ $\forall i,j = 1,2,...,n$. A reciprocal matrix is consistent if and only if $a_{ik} = a_{ij} \cdot a_{jk}$ $\forall i,j = 1,2,...,n$*

In the sequel we will consider A as a map $A : X \times X \to \mathbb{R}^+$ with $X = \{x_1, x_2, ..., x_n\}$ and $a_{ij} = A(x_i, x_j)$.

Consistent reciprocal matrices are characterized by the following theorem.

Theorem 7.38. [124] *An $n \times n$ real matrix A is reciprocal and consistent if and only if there exists a fuzzy subset μ of X such that*

$$a_{ij} = \frac{\mu(x_i)}{\mu_(x_j)} \; \forall i,j = 1,2,...,n.$$

The fuzzy subset μ will be called a generator of A and sometimes we will write A_μ to stress this fact.

Reciprocal consistent matrices generate betweenness relations.

Proposition 7.39. *Let A be a reciprocal consistent matrix on X with $a_{ij} \neq 1$ if $i \neq j$. A generates the following betweenness relation B_A on X.*

$$(x_i, x_j, x_k) \in B_A \text{ if and only if } a_{ik} = a_{ij} \cdot a_{jk} \text{ and } i \neq j \neq k \neq i.$$

Proposition 7.40. *Let A_μ be the reciprocal consistent matrix generated by the fuzzy subset μ of X and E_μ the T-indistinguishability operator on X generated by μ for T the product t-norm. The betweenness relations generated by A_μ and by E_μ coincide. (i.e. $B_{A_\mu} = B_{E_\mu}$).*

Proof. Let us order the elements of X in the following way: $x_i \leq x_j$ if and only if $\mu(x_i) \leq \mu(x_j)$. Then

$$\begin{aligned}
(x_i, x_j, x_k) \in B_{E_\mu} &\Leftrightarrow E(x_i, x_j) \cdot E(x_j, x_k) = E(x_i, x_k) \\
&\Leftrightarrow \frac{\mu(x_i)}{\mu(x_j)} \cdot \frac{\mu(x_j)}{\mu(x_k)} = \frac{\mu(x_i)}{\mu(x_k)} \\
&\Leftrightarrow a_{ij} \cdot a_{jk} = a_{ik} \\
&\Leftrightarrow (x_i, x_j, x_k) \in B_{A_\mu}.
\end{aligned}$$

For a given reciprocal matrix A, T.L Saaty obtains a consistent matrix A' close to A. A' is generated by an eigenvector associated to the greatest eigenvalue of A and fulfills the following properties.

- If A is already consistent, then $A = A'$.
- If A is a reciprocal positive matrix, then the sum of its eigenvalues is n.
- If A is consistent, then there exist a unique eigenvalue $\lambda_{\max} = n$ different from zero.
- Slight modifications of the entries of A produce slight changes to the entries of A'.

If a T-indistinguishability operator is close to a one dimensional one, then it almost generates a linear betweenness relation (the fuzzy betweenness relation that generates is close to a crisp linear one (see Section 6.6 of Chapter6). This suggests the following definition.

Definition 7.41. *Let T be the Product t-norm. Given $\epsilon \in [0,1]$, a T-indistinguishability operator E is ϵ-one dimensional if and only if there exists a total ordering $\leq_E$ in X such that for all $x, y, z \in X$ with $x \leq_E y \leq_E z$ the following inequality holds.*

$$|E(x,y) \cdot E(y,z) - E(x,z)| < \epsilon.$$

If E is an ϵ-one dimensional T-indistinguishability operator (T the Product t-norm) defined on a set X of cardinality n and $x_1 \leq_E x_2 \leq_E ... \leq_E x_n$ the total ordering on X determined by E, then the following definition provides a useful relation between reciprocal matrices and T-indistinguishability operators.

Definition 7.42. *The matrix A defined by*

$$A(x_i, x_j) = \begin{cases} E(x_i, x_j) & \text{if } i \leq j \\ \frac{1}{E(x_i, x_j)} & \text{otherwise} \end{cases}$$

will be called the ϵ-consistent reciprocal matrix associated to E.

Then in order to obtain a one dimensional T-indistinguishability operator E' close to a given one E (T the Product t-norm), the following **procedure** can be used:

- Find an $\epsilon > 0$ for which E is ϵ-one dimensional.
- Calculate the ϵ-consistent reciprocal matrix A associated to E.
- Find an eigenvector μ of the greatest eigenvalue of A.
- $E' = E_\mu$.

Example 7.43. Let T be the Product t-norm and E the T-indistinguishability operator on a set X of cardinality 5 given by the following matrix.

$$E = \begin{pmatrix} 1 & 0.74 & 0.67 & 0.50 & 0.41 \\ 0.74 & 1 & 0.87 & 0.65 & 0.53 \\ 0.67 & 0.87 & 1 & 0.74 & 0.60 \\ 0.50 & 0.65 & 0.74 & 1 & 0.80 \\ 0.41 & 0.53 & 0.60 & 0.80 & 1 \end{pmatrix}.$$

It is easy to verify that E is 0.1-one dimensional. Its associated 0.1-consistent reciprocal matrix A is

$$A = \begin{pmatrix} 1 & 1.3514 & 1.4925 & 2.0000 & 2.4390 \\ 0.7400 & 1 & 1.1494 & 1.5385 & 1.8868 \\ 0.6700 & 0.8700 & 1 & 1.3514 & 1.6667 \\ 0.5000 & 0.6500 & 0.7400 & 1 & 1.2500 \\ 0.4100 & 0.5300 & 0.6000 & 0.8000 & 1 \end{pmatrix}.$$

Its greatest eigenvalue is 5.0003 and an eigenvector for 5.0003 is

$$\mu = (1, 0.76, 0.67, 0.50, 0.40).$$

This fuzzy set generates E_μ which is a one dimensional T-indistinguishability operator close to E.

$$E_\mu = \begin{pmatrix} 1 & 0.76 & 0.67 & 0.50 & 0.40 \\ 0.76 & 1 & 0.88 & 0.66 & 0.53 \\ 0.67 & 0.88 & 1 & 0.74 & 0.60 \\ 0.50 & 0.66 & 0.74 & 1 & 0.81 \\ 0.40 & 0.53 & 0.60 & 0.81 & 1 \end{pmatrix}.$$

The results of this section can be easily generalized to continuous strict Archimedean t-norms.

If T' is a continuous strict Archimedean t-norm, then it is isomorphic to the Product t-norm T. Let f be this isomorphism. If E is a T'-indistinguishability operator, then $f \circ E$ is a T-indistinguishability operator. If $f \circ E$ is ϵ-one dimensional, then we can find E' one dimensional close to $f \circ E$ as before. Since isomorphisms between continuous t-norms are continuous and preserve dimensions, $f^{-1} \circ E'$ is a one dimensional T'-indistinguishability operator close to E.

8

Aggregation of Indistinguishability Operators

In many situations, there can be more than one indistinguishability operator or, more generally, a T-transitive relation defined on a universe. Let us suppose, for example that we have a set of instances defined by some features. We can generate an indistinguishability operator or a fuzzy preorder from each feature. Also, we can have some prototypes, and again we can define a relation from each of them in our universe. In these cases we may need to aggregate the relations obtained. This is usually done by calculating their minimum (or infimum). Although this has a very clear interpretation in fuzzy logic since the infimum is used to model the universal fuzzy quantifier $\forall$, it often leads to undesirable results in applications because the minimum has a drastic effect. If, for example, two objects of our universe are very similar or indistinguishable for all but one indistinguishability operator but are very different for this particular operator, then the application of the minimum will give this last measure all others will be forgotten. This can be reasonable and useful if we need a perfect matching with respect to all of our relations, but this is not the case in many situations. When we need to take all relations into account in a less dramatic way, we need other ways of aggregating them. Since if R and S are T-transitive fuzzy relations with respect to a t-norm T then $T(R, S)$ is also a T-transitive fuzzy relation, it seems at first glance that this could be a good way to aggregate them. Nevertheless, if we aggregate in this way, we obtain relations with very low values. In the case of non-strict Archimedean t-norms, it is even worse, since in many cases almost all of the values of the obtained relation are equal to zero. Therefore, other ways need to be found.

In this chapter, quasi-arithmetic means and a special type of OWA operators ([140], [142]) are used. Means and OWA operators can be thought of as fuzzy quantifiers between the universal $\forall$ and the existential $\exists$ ones ([90]), and they can use all the values of our relations in order to obtain their aggregation.

Section 8.1 will present maps that preserve transitivity in order to be able to aggregate transitive relations. Section 8.2 will focus on the use of

J. Recasens: Indistinguishability Operator, STUDFUZZ 260, pp. 147–161.
springerlink.com

quasi-arithmetic means to aggregate transitive relations. If t is an additive generator of a continuous Archimedean t-norm T, then the quasi-arithmetic mean of a T-transitive fuzzy relation with respect to the mean generated by t is also a T-transitive relation and does not depend on the selection of the generator t. This gives reasonable results, since the values of the obtained relations are a compromise between the values of the given ones and the mean is closely related to the t-norm used to model the transitivity.

Using these results, given a T-transitive fuzzy relation R three families $((R_\lambda)_{\lambda\in[0,1]}$, $({}^\lambda R)_{\lambda\in[0,1]})$ and $(R^\lambda)_{\lambda\in[0,1]})$ will be generated using the mean generated by t that will allow us to modify the values of R going from the smallest fuzzy relation to the largest one depending on the value of the parameter λ.

One interesting apect of aggregation using quasi-arithmetic means is that we can aggregate crisp relations and obtain fuzzy ones. This was first done by Bezdek and Harris in [10],[54], who obtained indistinguishability operators with respect to the Łukasiewicz t-norm from crisp equivalence relations using their weighted arithmetic means.

Section 8.3 will present another group of aggregation operators, sub additive OWA operators, for T-transitive fuzzy relations, when T is continuous Archimedean, which are independent of the additive generator t of T. In addition to their usefulness and popularity, these operators offer an interesting means of aggregation due to the relationship between OWA operators and quantifiers ([142]). Thanks to this we will have a nice semantic interpretation of the aggregation of T-transitive fuzzy relations using OWA operators.

In Section 8.4, the results will be extended to aggregate a non-finite family of transitive relations. The obtained results will be applied to calculate the degree of inclusion and similarity of fuzzy quantities (fuzzy subsets of an interval of the real line).

8.1 Aggregating T-Transitive Fuzzy Relations

The most common way to aggregate a family of T-transitive fuzzy relations is calculating their infimum, which also is a T-transitive relation (Proposition 2.12). This means that a couple (x, y) are related with respect to this relation if and only if they are related with respect to all the relations of the family since the infimum is used to model the universal quantifier $\forall$ in fuzzy logic [54].

Nevertheless, in many situations this way of aggregating fuzzy relations leads to undesirable results since the Infimum only takes the smaller value for every couple into account and forgets or loses the information of the other values.

A possibility to soften the previous proposition is replacing the Infimum by the t-norm T as the following proposition shows.

Proposition 8.1. *Let $R_1, R_2, ...R_n$ be n T-transitive fuzzy relations on a set X. The relation R defined for all $x, y \in X$ by*

$$R(x, y) = T(R_1(x, y), R_2(x, y), ..., R_n(x, y))$$

is a T-transitive fuzzy relation on X.

Proof. $\forall i = 1, 2, ..., n$, $T(R_i(x, y), R_i(y, z)) \leq R_i(x, z)$.

Since T is non decreasing in both variables,

$$T(T(R_1(x, y), R_1(y, z)), T(R_2(x, y), R_2(y, z)), ..., T(R_n(x, y), R_n(y, z)))$$
$$\leq T(R_1(x, z), R_2(x, z), ..., R_n(x, z))$$

and due to the associativity and commutativity of T,

$$T(T(R_1(x, y), R_2(x, y), ..., R_n(x, y)), T(R_1(y, z), R_2(y, z), ...R_n(y, z)))$$
$$\leq T(R_1(x, z), R_2(x, z), ..., R_n(x, z)).$$

If T is assumed to model the 'and' connective, instead of using the fuzzy quantifier $\forall$ (i.e. taking the infimum) we are saying that two elements x and y are related by R if and only if they are related by R_1 and R_2... and R_n. This way to aggregate relations has some advantages with respect to using the minimum or infimum. For example, if we want to aggregate two relations and one is twice as important as the other, we can count the most important twice. But on the other hand, it can produce relations with very small values and if the t-norm is non-strict Archimedean most of them will probably be 0.

So, more general ways to aggregate transitive relations are needed.

Definition 8.2. [113] *Given a family $\mathcal{T} = (T_i)_{i=1,2,...,n}$ of t-norms and a t-norm T, a function $f : [0, 1]^n \to [0, 1]$ is an operator that aggregates transitive fuzzy relations with respect to $\mathcal{T}$ and T when, for any non-empty set X and any arbitrary collection $(R_i)_{i=1,2,...,n}$ of fuzzy relations on X with R_i T_i-transitive for all $i = 1, 2, ..., n$, the fuzzy relation $R = f(R_1, R_2, ..., R_n)$ on X is T-transitive.*

Proposition 8.3 shows that in the continuous Archimedean case f can be written in the form $f = t^{[-1]} \circ s \circ (t_1 \times t_2 \times ... \times t_n)$ with $s : (\mathbb{R}^+)^n \to \mathbb{R}^+$ a sub-additive map while Proposition 8.4 gives a sufficient condition for an operator f to aggregate transitive relations. They generalize the results of [113] (see also Proposition 4.5).

Proposition 8.3. *Let $f : [0, 1]^n \to [0, 1]$ be an operator that aggregates transitive fuzzy relations with respect to $\mathcal{T} = (T_i)_{i=1,2,...,n}$ and T where T and the members of $\mathcal{T}$ are continuous Archimedean t-norms with generators $t, t_1, ..., t_n$ respectively. Then $f = t^{[-1]} \circ s \circ (t_1 \times t_2 \times ... \times t_n)$ with $s : (\mathbb{R}^+)^n \to \mathbb{R}^+$ a sub-additive map.*

Proof. The proof is very similar to the proof of Proposition 4.5.

For any transitive fuzzy relation R_i with respect to T_i, let $R_i(x,y) = a_i$, $R_i(y,z) = b_i$ and $R_i(x,z) = c_i$ and $t(a_i) = u_i$, $t(b_i) = v_i$, $t(c_i) = w_i$. Since f aggregates transitive fuzzy relations,

$$T(f(a_1, a_2, ..., a_n), f(b_1, b_2, ..., b_n)) \leq f(c_1, c_2, ..., c_n)$$

or

$$t(f(a_1, a_2, ..., a_n)) + t(f(b_1, b_2, ..., b_n)) \geq t(f(c_1, c_2, ..., c_n))$$

and

$$t(f(t_1^{[-1]}(u_1), t_2^{[-1]}(u_2), ..., t_n^{[-1]}(u_n))) + t(f(t_1^{[-1]}(v_1), t_2^{[-1]}(v_2), ..., t_n^{[-1]}(v_n)))$$
$$\geq t(f(t_1^{[-1]}(w_1), t_2^{[-1]}(w_2), ..., t_n^{[-1]}(w_n))).$$

Putting $s = t \circ f \circ (t_1^{[-1]} \times t_2^{[-1]} \times ... \times t_n^{[-1]})$, s is sub-additive and $f = t^{[-1]} \circ s \circ (t_1 \times t_2 \times ... \times t_n)$.

Proposition 8.4. *Let $f : [0,1]^n \to [0,1]$ be a map and $T, T_1, T_2, ..., T_n$ continuous Archimedean t-norms with additive generators $t, t_1, t_2, ..., t_n$ respectively. If there exists a non decreasing sub-additive map $s : (\mathbb{R}^+)^n \to \mathbb{R}^+$ such that $f = t^{[-1]} \circ s \circ (t_1 \times t_2 \times ... \times t_n)$, then f aggregates transitive relations.*

Proof. The proof is very similar to the proof of Proposition 4.5.

Corollary 8.5. [113] *Let $f : [0,1]^n \to [0,1]$ be a map and $T, T_1, T_2, ..., T_n$ continuous Archimedean t-norms with additive generators $t, t_1, t_2, ..., t_n$ respectively. If there exists a metric transform $s : (\mathbb{R}^+)^n \to \mathbb{R}^+$ such that $f = t^{[-1]} \circ s \circ (t_1 \times t_2 \times ... \times t_n)$, then f aggregates indistinguishability operators and fuzzy preorders.*

Proof. Since s is a metric transform, reflexivity is also assured and symmetry is trivially preserved.

As it is pointed out in [113],[114], this way to aggregate relations is not canonical, in the sense that it depends in general on the particular selection of the additive generators of the t-norms. This means that aggregating the same relations using the same sub-additive map s we can obtain different results, which does not seem a very intuitive property for aggregating relations. Let us investigate when the obtained relation does not depend on the selection of the generator (i.e. it is canonical) in the case when all the t-norms coincide.

Proposition 8.6. *Let $f = t^{[-1]} \circ s \circ (t \times t \times ... \times t)$ be a map that aggregates n T-transitive relations ($s : (\mathbb{R}^+)^n \to \mathbb{R}^+$). f does not depend on the particular generator t of T if and only if $\forall \alpha > 0$ and $\forall \vec{x} \in (\mathbb{R}^+)^n$ $s(\alpha \vec{x}) = \alpha s(\vec{x})$.*

Proof

$\Rightarrow$)

Two additive generators t and t' of T differ by a positive multiplicative constant α. Therefore $t'(x) = \alpha t(x)$ and $t'^{[-1]}(x) = t^{[-1]}(\frac{y}{\alpha})$.

Let $(x_1, x_2, ..., x_n) \in [0,1]^n$

If $t^{[-1]}(s(t(x_1), t(x_2), ..., t(x_n))) = t^{[-1]}(\frac{s(\alpha(t(x_1),t(x_2),...,t(x_n))}{\alpha})$, then

$$\alpha s((t(x_1), t(x_2), ..., t(x_n))) = s(\alpha t((t(x_1), t(x_2), ..., t(x_n)))$$

and putting $(t(x_1), t(x_2), ..., t(x_n)) = \vec{y}$,

$$\alpha s(\vec{y}) = s(\alpha \vec{y}) \ \forall \alpha > 0, \forall \vec{y} \in [0, t(0)]^n.$$

Since for every $(x_1, x_2, ..., x_n) \in (\mathbb{R}^+)^n$ we can find a generator t with $t(0) > x_i$ $i = 1, 2, ..., n$, $s(\alpha \vec{x}) = \alpha s(\vec{x})$ in all $(\mathbb{R}^+)^n$.

$\Leftarrow$)

Trivial.

Corollary 8.7. *Let $f = t^{[-1]} \circ s \circ (t \times t \times ... \times t)$ be a map that aggregates n T-transitive relations ($s : (\mathbb{R}^+)^n \to \mathbb{R}^+$) with s a linear map. Then f does not depend on the particular generator t of T.*

For $n = 1$ we get the following result.

Corollary 8.8. *The only canonical preserving transitivity operators are the automorphisms of the continuous Archimedean t-norm T.*

Proof. If $n = 1$ in the last proposition, then s is a linear map $s(x) = \alpha x$ and $f = t^{[-1]}\alpha t$. (cf. Propositions 4.5 and 4.10).

Finally, the operators aggregating min-transitive fuzzy relations are non decreasing maps.

Proposition 8.9. [113], [114] *A map $f : [0,1]^n \to [0,1]$ aggregates* min-*transitive fuzzy relations if and only if f is a non decreasing map.*

8.2 Aggregating T-Transitive Fuzzy Relations Using Quasi-arithmetic Means

From Corollary 8.7 an operator $f : [0,1]^n \to [0,1]$ preserving T-transitivity of the form $f = t^{[-1]} \circ s \circ t$ with s a linear map is canonical. If $s(x_1, x_2, ...x_n) = p_1 x_1 + p_2 x_2 + ... + p_n x_n$ then f has the form

$$f(x_1, x_2, ..., x_n) = t^{[-1]}(p_1 t(x_1) + p_2 t(x_2) + ... + p_n t(x_n))) \text{ with } p_1, p_2, ..., p_n > 0.$$

If we take $s(x_1, x_2, ..., x_n) = x_1 + x_2 + ... + x_n$ (i.e.: $p_i = 1$ for all $i = 1, 2, ..., n$), then we aggregate using the t-norm T and recover Proposition 8.1.

Taking $s(x_1, x_2, ..., x_n) = (x_1 + x_2 + ... + x_n)/n$ (i.e.: $p_i = 1/n$ for all $i = 1, 2, ..., n$), then we get arithmetic means, and choosing $s(x_1, x_2, ..., x_n) = p_1 x_1 + p_2 x_2 + ... + p_n x_n$ with $\sum_{i=1}^n p_i = 1$ we obtain weighted means.

So means are canonical operators preserving T-transitivity and from a family of relations they give as a result a relation with values between the greatest and the lowest ones of the family. Therefore they seem a good way to aggregate T-transitive relations.

Since the additive generator of a continuous Archimedean t-norm is a decreasing map $t : [0, 1] \to \mathbb{R}$, this gives a way to generate a quasi-arithmetic mean m_t associated to a continuous Archimedean t-norm. Next we will show that this gives is a natural bijection between continuous Archimedean t-norms and quasi-arithmetic means.

Lemma 8.10. *Let $t, t' : [0, 1] \to \mathbb{R}$ be two continuous strict monotonic maps differing only by an additive constant. Then $m_t = m_{t'}$.*

Proof. If $t' = t + a$, then

$$
\begin{aligned}
m_{t'}(x, y) &= t'^{-1}(\frac{t'(x) + t'(y)}{2}) \\
&= t^{-1}(\frac{t(x) + a + t(y) + a}{2} - a) \\
&= t^{-1}(\frac{t(x) + t(y)}{2}) = m_t(x, y).
\end{aligned}
$$

Lemma 8.11. *Let $t : [0, 1] \to \mathbb{R}$ be a continuous strict monotonic map. Then $m_t = m_{-t}$.*

Proof

$$
\begin{aligned}
m_{-t}(x, y) &= (-t)^{-1}(\frac{(-t)(x) + (-t)(y)}{2}) \\
&= t^{-1}(-\frac{-t(x) - t(y)}{2}) \\
&= t^{-1}(\frac{t(x) + t(y)}{2}) = m_t(x, y).
\end{aligned}
$$

Lemma 8.12. *Let $t : [0, 1] \to \mathbb{R}$ be a continuous strict monotonic map and $\alpha > 0$. Then $m_t = m_{\alpha t}$.*

Proof

$$
\begin{aligned}
m_{\alpha t}(x, y) &= (\alpha t)^{-1}(\frac{(\alpha t)(x) + (\alpha t)(y)}{2}) \\
&= t^{-1}(\frac{\alpha t(x) + \alpha t(y)}{2\alpha}) \\
&= t^{-1}(\frac{t(x) + t(y)}{2}) = m_t(x, y).
\end{aligned}
$$

As a consequence of these three lemmas the following result follows

Proposition 8.13. *The map assigning to every continuous Archimedean t-norm T with additive generator t the quasi-arithmetic mean m_t generated by t is a canonical bijection between the set of continuous Archimedean t-norms and continuous quasi-arithmetic means.*

It is straightforward to extend the previous proposition to more than two variables and to weighted quasi-arithmetic means.

Weighted quasi-arithmetic means are defined in the following way.

Definition 8.14. *Let $p_1, p_2, ..., p_n$ be positive numbers such that $\sum_{i=1}^n p_i = 1$. p_i are called weights. The quasi-arithmetic mean of $x_1, x_2, ..., x_n \in [0,1]$ with weights $p_1, p_2, ..., p_n$ generated by t, a continuous strict monotonic map $t : [0,1] \to [-\infty, \infty]$, is*

$$m_t(x,y) = t^{-1}\left(\frac{\sum_{i=1}^n p_i \cdot t(x_i)}{n}\right).$$

By the use of weighted quasi-arithmetic means, if we have two (or more) T-transitive fuzzy relations R and S, we can create families of T-transitive fuzzy relations depending on one or more parameters allowing us to go from R to S. This gives interesting results if one of them is a crisp special one. We will study the cases when one of them is the crisp equality Id, the smallest relation $\mathbf{0}$ with all entries equal to 0 or the greatest one $\mathbf{1}$ with all entries equal to 1 in our universe. This means to consider the following families $(R_\lambda)_{\lambda\in[0,1]}$, $({}^\lambda R)_{\lambda\in[0,1]}$ and $(R^\lambda)_{\lambda\in[0,1]}$ for a given T-transitive fuzzy relation R on a set X:

Definition 8.15. *Let R be a T-transitive fuzzy relation on a set X with T a continuous Archimedean t-norm and t an additive generator of T. The families $(R_\lambda)_{\lambda\in[0,1]}$, $({}^\lambda R)_{\lambda\in[0,1]}$ and $(R^\lambda)_{\lambda\in[0,1]}$ are defined by*

$$\begin{aligned} R_\lambda &= t^{[-1]}(\lambda t(R) + (1-\lambda)t(Id)) \\ {}^\lambda R &= t^{[-1]}(\lambda t(R) + (1-\lambda)t(\mathbf{0})) \\ R^\lambda &= t^{[-1]}(\lambda t(R) + (1-\lambda)t(\mathbf{1})) = t^{[-1]}(\lambda t(R)). \end{aligned}$$

a) $(R^\lambda)_{\lambda\in[0,1]}$ goes from R when $\lambda = 1$ to $\mathbf{1}$ when $\lambda = 0$.
b) $(R_\lambda)_{\lambda\in[0,1]}$ goes from R when $\lambda = 1$ to Id when $\lambda = 0$ if T is a non-strict Archimedean t-norm. If T is strict, then $R_\lambda(x,y) = 0$ if $x \neq y$ and $R_\lambda(x,x) = t^{-1}(\lambda t(R(x,x))$ $\forall \lambda \in [0,1)$ and we define $R_1 = R$.
c) If T is non-strict, then $({}^\lambda R)_{\lambda\in[0,1]}$ goes from R when $\lambda = 1$ to $\mathbf{0}$ when $\lambda = 0$. If T is strict, then ${}^\lambda R = \mathbf{0}$ $\forall \lambda \in [0,1)$ and we define ${}^1 R = R$.

Example 8.16. a) If R is a fuzzy relation transitive with respect to the Łukasiewicz t-norm, then

$$R_\lambda(x,y) = \begin{cases} \lambda R(x,y) & \text{if } x \neq y \\ \lambda R(x,y) + 1 - \lambda & \text{if } x = y. \end{cases}$$

$$R^\lambda(x,y) = \lambda R(x,y) + 1 - \lambda$$

$${}^\lambda R(x,y) = \lambda R(x,y).$$

b) If R is a fuzzy relation transitive with respect to the Product t-norm, then

$$R^\lambda(x,y) = (R(x,y))^\lambda.$$

These families can be combined to obtain $(R_{\lambda,\mu,\rho})_{\lambda,\mu,\rho\in[0,1]}$:

$$R_{\lambda,\mu,\rho} = t^{[-1]}((1-\lambda-\mu-\rho)t(R) + \lambda t(Id) + \mu t(\mathbf{0}) + \rho t(\mathbf{1}))$$

Example 8.17. If R is a fuzzy relation transitive with respect to the Łukasiewicz t-norm T, then

$$R_{\lambda,\mu,\rho} = \begin{cases} (1-\lambda-\mu-\rho)R(x,y) + \rho & \text{if } x \neq y \\ (1-\lambda-\mu-\rho)R(x,y) + \rho + \lambda & \text{if } x = y. \end{cases}$$

For $\lambda = 0$ we obtain that a linear change of a T-transitive relation $R \to \alpha R + \beta$ with $\alpha + \beta = 1$ produces a T-transitive relation.

If R is a one dimensional T-indistinguishability operator or fuzzy T-preorder, we obtain the following results.

Proposition 8.18. *Let P_μ be the one dimensional fuzzy T-preorder on a set X with T the Product t-norm generated by the fuzzy subset μ of X. Then $(P_\mu)^\lambda = P_{\mu^\lambda}$ (i.e.: $(P_\mu)^\lambda$ is the one dimensional fuzzy T-preorder generated by the fuzzy subset μ^λ of X).*

Proof

$$\begin{aligned}(P_\mu)^\lambda(x,y) &= (P_\mu(x,y))^\lambda = \min(1, \frac{\mu(y)}{\mu(x)})^\lambda \\ &= \min(1, \frac{\mu(y)^\lambda}{\mu(x)^\lambda}) = P_{\mu^\lambda}(x,y).\end{aligned}$$

Similarly,

Proposition 8.19. *Let E_μ be the one dimensional fuzzy T-indistinguishability operator on a set X with T the Product t-norm generated by the fuzzy subset μ of X. Then $(E_\mu)^\lambda = E_{\mu^\lambda}$ (i.e.: $(E_\mu)^\lambda$ is the one dimensional fuzzy T-indistinguishability operator generated by μ^λ).*

Proof

$$(E_\mu)^\lambda(x,y) = E_\mu(x,y)^\lambda = \min(\frac{\mu(x)}{\mu(y)}, \frac{\mu(y)}{\mu(x)})^\lambda$$
$$= \min(\frac{\mu(x)^\lambda}{\mu(y)^\lambda}, \frac{\mu(y)^\lambda}{\mu(x)^\lambda}) = E_{\mu^\lambda}(x,y).$$

It is worth noticing that there are no similar results to the two previous ones for non-strict continuous Archimedean t-norms.

Aggregating using quasi-arithmetic means and weighted means allow us to obtain fuzzy relations from crisp ones. This was first done by Bezdek and Harris, who calculated the weighted arithmetic means of crisp equivalence relations in order to obtain hard partitions [10]. Their results can be reinterpreted in this more general framework and extended so that transitive crisp relations can be aggregated in order to obtain T-transitive relations with respect to general non-strict continuous Archimedean t-norms.

8.3 Aggregating *T*-Transitive Fuzzy Relations Using OWA Operators

OWA operators were introduced by Yager ([140]) and have been used in many applications where there is a need to joint the information contained in more than one fuzzy set. There are in general operators related to the weighted arithmetic mean and therefore to the Łukasiewicz t-norm, but as we will see in this section, the same idea can be generalized to any continuous Archimedean t-norm.

OWA operators are very interesting non linear solutions of the functional equation $\alpha s(\vec{x}) = s(\alpha \vec{x})$. Therefore, thanks to Proposition 8.6 some of them could be suitable to aggregate T-transitive fuzzy relations (for T a continuous Archimedean t-norm) since the result would be independent of the generator of the t-norm used (i.e. they could generate canonical aggregation operators). Actually, we will prove that sub-additive OWA operators are.

Definition 8.20. *An OWA operator is a mapping $s : \mathbb{R}^n \to \mathbb{R}$ that has associated a vector of weights $(p_1, p_2, ..., p_n)$ such that $p_i \in [0,1]$ $\forall i = 1,2,...,n$ and $\sum_{i=1}^n p_i = 1$ where $s(x_1, x_2, ..., x_n) = \sum_{i=1}^n p_i x_{(i)}$ where $x_{(i)}$ is the ith largest of the x_i.*

Not all OWA operator will aggregate T-transitive fuzzy relations since as proved in Proposition 8.3 they have to be sub-additive maps. Since OWA operators are non decreasing maps, thanks to Proposition 8.4 this will be a sufficient condition as well.

Proposition 8.21. *Let $s : \mathbb{R}^n \to \mathbb{R}$ be the OWA operator with weights $(p_1, p_2, ..., p_n)$. s is a sub-additive map if and only if $p_i \geq p_j$ for $i < j$.*

Proof

$\Rightarrow$) Let us fix $i < n$. For every a and b with $a > b$ we can consider the vectors $\vec{x} = (x_1, x_2, ..., x_n)$ and $\vec{y} = (y_1, y_2, ..., y_n)$ defined by

$$x_1 = x_2 = ... = x_i = a, x_{i+1} = ... = x_n = b$$

and

$$y_1 = y_2 = ... = y_{i-1} = a, y_i = b, y_{i+1} = a, y_{i+2}, = ... = y_n = b.$$

$s(\vec{x} + \vec{y})$ and $s(\vec{x}) + s(\vec{y})$ differ only in the ith and the $(i+1)^{\text{th}}$ coordinates. The difference is

$$(a+b)p_i + (a+b)p_{i+1} - 2ap_i + 2bp_{i+1}.$$

If s is sub-additive, then

$$(a+b)p_i + (a+b)p_{i+1} \leq 2ap_i + 2bp_{i+1} \ \forall a > b$$

Putting $c = a - b$, this is equivalent to

$$2bp_i + cp_i + 2bp_{i+1} + cp_{i+1} \leq 2bp_i + 2cp_i + 2bp_{i+1}$$

which is satisfied if and only if $p_i \geq p_{i+1}$.

$\Leftarrow$) Trivial.

As a corollary we obtain the following proposition which characterizes the operators that aggregate T-transitive relations generated by OWA operators:

Proposition 8.22. *Let $s : \mathbb{R}^n \to \mathbb{R}$ be the OWA operator with weights $(p_1, p_2, ..., p_n)$ and t an additive generator of a continuous Archimedean t-norm T. $f = t^{[-1]} \circ s \circ t$ aggregates T-transitive fuzzy relations if and only if $p_i \geq p_j$ for $i < j$.*

It is interesting to note that since t is a decreasing map, we are giving more importance to small values and in this sense they are aggregations softening the aggregation using the minimum. In fact, the minimum is achieved taking $p_1 = 1$ and $p_i = 0 \ \forall i = 2, ..., n$ and this gives an alternative proof to the fact that the minimum of T-transitive relations is also a T-transitive relation.

Example 8.23. Let us consider the two T-indistinguishability operators E_1 and E_2 on $X = \{x, y, z\}$ (T the Łukasiewicz t-norm) defined by the following matrices:

$$E_1 = \begin{matrix} & \begin{matrix} x & y & z \end{matrix} \\ \begin{matrix} x \\ y \\ z \end{matrix} & \begin{pmatrix} 1 & 0.2 & 0.8 \\ 0.2 & 1 & 0.4 \\ 0.8 & 0.4 & 1 \end{pmatrix} \end{matrix}$$

$$E_2 = \begin{array}{c} \\ x \\ y \\ z \end{array} \begin{array}{c} \begin{array}{ccc} x & y & z \end{array} \\ \begin{pmatrix} 1 & 0.6 & 0.4 \\ 0.6 & 1 & 0.8 \\ 0.4 & 0.8 & 1 \end{pmatrix} \end{array}$$

If we consider the OWA operator with parameters $p_1 = 3/4$ and $p_2 = 1/4$, we obtain the T-indistinguishability operator E with matrix

$$E = \begin{pmatrix} 1 & 0.3 & 0.5 \\ 0.3 & 1 & 0.5 \\ 0.5 & 0.5 & 1 \end{pmatrix}.$$

The arithmetic mean of E_1 and E_2 is

$$m(E_1, E_2) = \begin{pmatrix} 1 & 0.4 & 0.6 \\ 0.4 & 1 & 0.6 \\ 0.6 & 0.6 & 1 \end{pmatrix},$$

the minimum is

$$\min(E_1, E_2) = \begin{pmatrix} 1 & 0.2 & 0.4 \\ 0.2 & 1 & 0.4 \\ 0.4 & 0.4 & 1 \end{pmatrix}$$

and the aggregation using T is

$$T(E_1, E_2) = \begin{pmatrix} 1 & 0 & 0.2 \\ 0 & 1 & 0.2 \\ 0.2 & 0.2 & 1 \end{pmatrix}.$$

8.4 Aggregating a Non-finite Number of T-Transitive Fuzzy Relations

In some cases we have to aggregate a non-finite number of relations. Suppose that we have a family of T-transitive relations $(R_i)_{i \in [a,b]}$ on a set X with the indices in the interval $[a, b]$ of the real line and that for every couple (x, y) of X the map $f_{(x,y)} : [a, b] \to \mathbb{R}$ defined by $f_{(x,y)}(i) = R_i(x, y)$ is integrable in some sense. Then we can replace the map s used to aggregate the fuzzy relation in the finite case by a map related to the integration of the family.

Definition 8.24. *With the previous notations, the aggregation of the family $(R_i)_{i \in [a,b]}$ with respect to a continuous Archimedean t-norm T with additive generator t is the fuzzy relation R defined for all $x, y \in X$ by*

$$R(x, y) = t^{[-1]}(\int_a^b t(R_i(x, y))di)$$

Proposition 8.25. *This definition is independent of the generator of T.*

Proof. If $t' = \alpha t$ is another additive generator of T, then

$$\begin{aligned} & t'^{[-1]}(\int_a^b t'(R_i(x,y))di) \\ & = t^{[-1]}(\frac{\int_a^b \alpha t(R_i(x,y))di}{\alpha}) \\ & = t^{[-1]}(\int_a^b t(R_i(x,y))di). \end{aligned}$$

Definition 8.24 is the continuous version of Proposition 8.1 and it allow us to aggregate continuous families of T-transitive fuzzy relations using the given t-norm T. With more reason than in the finite case, we will obtain very low values for $R(x,y)$ in general and if the t-norm is non-strict, then it is most likely that most of them be zero. Therefore, we need to find an aggregation generalizing the quasi-arithmetic mean of the finite case:

Definition 8.26. *With the previous notations, the mean aggregation of the family* $(R_i)_{i\in[a,b]}$ *with respect to* T *is the fuzzy relation* R *defined for all* $x,y \in X$ *by*

$$R(x,y) = t^{[-1]}(\frac{1}{b-a}\int_a^b t(R_i(x,y))di)$$

This definition is also independent of the generator of T, thanks to the linearity of integration.

Probably the most important necessity of aggregating a non finite family of fuzzy relations is when we need to compare two fuzzy subsets μ and ν of our universe X (i.e.: calculating their degree of similarity or indistinguishability or calculating the degree in which one is contained in the other.)

Again, one of the most popular ways is to compare $\mu(x)$ and $\nu(x)$ for all $x \in X$ using E_T or $\overrightarrow{T}$ and then taking the infimum of all the results.

Definition 8.27. *Let* μ, ν *be two fuzzy subsets of a set* X *and* T *a t-norm. The degree of similarity* $E_T(\mu,\nu)$ *between* μ *and* ν *is defined by*

$$E_T(\mu,\nu) = \inf_{x\in X} E_T(\mu(x),\nu(x)).$$

(cf. Definition 3.79).

This definition fuzzifies the crisp equality between two subsets A and B of X:

$$A = B \text{ if and only if } \forall x \in X \; x \in A \Leftrightarrow \; x \in B.$$

Definition 8.28. *Let* μ, ν *be two fuzzy subsets of a set* X. *The degree of inclusion* $P_T(\mu,\nu)$ *of* μ *into* ν *is defined by*

$$P_T(\mu,\nu) = \inf_{x\in X} \overrightarrow{T}(\mu(x),\nu(x)).$$

This definition fuzzifies the inclusion of a crisp subset A into a crisp subset B of X:

$$A \subseteq B \text{ if and only if } \forall x \in X \ x \in A \Rightarrow \ x \in B.$$

Lemma 8.29. *Given a t-norm T and an element x of a set X, the relation P_T^x defined on the set of fuzzy subsets of X by*

$$P_T^x(\mu,\nu) = \overrightarrow{T}(\mu(x)|\nu(x)).$$

is a T-preorder.

Corollary 8.30. *Given a t-norm T and a set X, the relation P_T defined on the set of fuzzy subsets of X is a T-preorder.*

Both definitions suffer from the drastic effect of the infimum. For example, if we have two fuzzy subsets μ,ν of a set X with $\mu(x)=\nu(x)$ for all $x\in X$ except for a value x_0 for which $\mu(x_0)=1$ and $\nu(x_0)=0$, then $E_T(\mu,\nu)=0$ and $P_T(\mu,\nu)=0$ which means that both subsets are considered completely different or dissimilar and that μ can not be seen contained in ν at any degree. An average of the values obtained for every $x\in X$ seems a suitable alternative.

Definition 8.31. *Let μ,ν be two integrable fuzzy subsets of an interval $[a,b]$ of the real line. The averaging degree of similarity or indistinguishability $E_T^A(\mu,\nu)$ between μ and ν with respect to a continuous Archimedean t-norm T with additive generator t is defined by*

$$E_T^A(\mu,\nu) = t^{[-1]}\left(\frac{1}{b-a}\int_a^b t(E_T(\mu(x),\nu(x)))dx\right).$$

Proposition 8.32. *The fuzzy relation E_T^A defined on the set of integrable fuzzy subsets $\mathcal{F}I_{[a,b]}$ of an interval $[a,b]$ of the real line is a T-indistinguishability operator and does not depend on the selection of the additive generator t of the t-norm.*

Proof. It is trivial to prove that E_T^A is a reflexive and symmetric fuzzy relation.

Let us prove that it is T-transitive.

Let $\mu,\nu,\rho\in\mathcal{F}I_{[a,b]}$.

$$\begin{aligned}
&T(E_T^A(\mu,\rho),E_T^A(\rho,\nu)\\
&= t^{[-1]}(t\circ t^{[-1]}(\frac{1}{b-a}\int_a^b t(E_T(\mu(x),\rho(x)))dx)\\
&+ t\circ t^{[-1]}(\frac{1}{b-a}\int_a^b t(E_T(\rho(x),\nu(x)))dx))
\end{aligned}$$

$$= t^{[-1]}(\frac{1}{b-a}\int_a^b t(E_T(\mu(x),\rho(x)))dx + \frac{1}{b-a}\int_a^b t(E_(\rho(x),\nu(x)))dx)$$

$$= t^{[-1]}(\frac{1}{b-a}\int_a^b (t(E_T(\mu(x),\rho(x))) + t(E_T(\rho(x),\nu(x))))dx).$$

$$t(E_T(\mu(x),\rho(x))) + t(E_T(\rho(x),\nu(x))) \geq t(E_T(\mu(x),\nu(x))) \; \forall x \in X$$

and therefore

$$\int_a^b (t(E_T(\mu(x),\rho(x))) + t(E_T(\rho(x),\nu(x))))dx \geq \int_a^b (t(E_T(\mu(x),\nu(x))))dx$$

So, since $t^{[-1]}$ is a non increasing map, we get

$$T(E_T^A(\mu,\rho), E_T^A(\rho,\nu) \leq E_T^A(\mu,\nu).$$

Definition 8.33. *Let μ, ν be two integrable fuzzy subsets of an interval $[a,b]$ of the real line. The averaging degree of inclusion $P_T^A(\mu,\nu)$ of μ into ν is defined by*

$$P_T^A(\mu,\nu) = t^{[-1]}\left(\frac{1}{b-a}\int_a^b t(P_T^x(\mu,\nu))dx\right).$$

or equivalently

$$P_T^A(\mu,\nu) = t^{[-1]}\left(\frac{1}{b-a}\int_a^b t(\overrightarrow{T}(\mu(x)|\nu(x)))dx\right).$$

Proposition 8.34. *The fuzzy relation P_T^A defined on the set of integrable fuzzy subsets of an interval $[a,b]$ of the real line is a fuzzy T-preorder and does not depend on the selection of the additive generator t of the t-norm.*

With these definitions, the degrees of indistinguishability and inclusion of the two fuzzy subsets after Corollary 8.30 are 1, which is a very intuitive result.

Example 8.35. Let us consider the two fuzzy subsets μ and ν of the interval $[0,2]$ defined by $\mu(x) = 1/2$ and $\nu(x) = x/2 \; \forall x \in [0,2]$. Let T_α be the Yager family of t-norms ($T_\alpha(x,y) = 1 - \min(1,(1-x)^\alpha + (1-y)^\alpha)^{\frac{1}{\alpha}}$ and $t_\alpha(x) = (1-x)^\alpha$ a generator of T_α with $\alpha \in (0,\infty)$).
a)

$$\overrightarrow{T}_\alpha(x|y) = \min(1-((1-y)^\alpha - (1-x)^\alpha)^{\frac{1}{\alpha}}, 1)$$

which implies that

$$P_{T_\alpha}^x(\mu,\nu)) = \min(1-((1-\nu(x))^\alpha - (1-\mu(x))^\alpha)^{\frac{1}{\alpha}}), 1)$$
$$= \begin{cases} 1 & \text{if } x > 1 \\ 1-((1-\frac{x}{2})^\alpha - (\frac{1}{2})^\alpha)^{\frac{1}{\alpha}} & \text{if } x \leq 1. \end{cases}$$

and therefore

$$P_T^A(\mu,\nu) = t^{[-1]}(\frac{1}{2}(\int_0^1 ((1-\frac{x}{2})^\alpha - (\frac{1}{2})^\alpha)dx) + \int_1^2 t(1)dx))$$
$$= t^{-1}(-\frac{(\frac{1}{2})^{\alpha+1}}{\alpha+1} - (\frac{1}{2})^{\alpha+1} + \frac{1}{\alpha+1})$$
$$= 1 - (-\frac{(\frac{1}{2})^{\alpha+1}}{\alpha+1} - (\frac{1}{2})^{\alpha+1} + \frac{1}{\alpha+1})^{\frac{1}{\alpha}}.$$

b)

$$E_{T_\alpha}(x,y) = 1 - |(1-y)^\alpha - (1-x)^\alpha|^{\frac{1}{\alpha}}$$

which implies that

$$E_{T_\alpha}^x(\mu,\nu)) = 1 - |(1-\nu(x))^\alpha - (1-\mu(x))^\alpha|^{\frac{1}{\alpha}}$$
$$= \begin{cases} 1-((\frac{1}{2})^\alpha - 1 - \frac{x}{2})^\alpha)^{\frac{1}{\alpha}} & \text{if } x > 1 \\ 1-((1-\frac{x}{2})^\alpha - (\frac{1}{2})^\alpha)^{\frac{1}{\alpha}} & \text{if } x \leq 1. \end{cases}$$

and therefore

$$E_{T_\alpha}^A(\mu,\nu) = t^{[-1]}(\frac{1}{2}(\int_0^1 ((1-\frac{x}{2})^\alpha - (\frac{1}{2})^\alpha)dx) + \int_1^2 ((\frac{1}{2})^\alpha - (1-\frac{x}{2})^\alpha)dx))$$
$$= t^{-1}(\frac{1-(\frac{1}{2})^\alpha}{\alpha+1}) = 1 - (\frac{1-(\frac{1}{2})^\alpha}{\alpha+1})^{\frac{1}{\alpha}}.$$

If $\alpha = 1$, then T_α is the Łukasiewicz t-norm T and in this case the previous formulas give $P_T^A(\mu,\nu) = 7/8$ and $E_T^A(\mu,\nu) = 3/4$ whereas using the infimum to aggregate we obtain $P_T(\mu,\nu) = E_T(\mu,\nu) = 1/2$.

It is interesting to notice that when $\alpha \to \infty$ then the Yager family tends to the minimum t-norm. If we calculate the limits of $P_T^A(\mu,\nu)$ and $E_T^A(\mu,\nu)$ when $\alpha \to \infty$ we obtain $\lim_{\alpha\to\infty} P_T^A(\mu,\nu) = \lim_{\alpha\to\infty} E_T^A(\mu,\nu) = 0$ which coincide with $P_{\min}(\mu,\nu)$ and $E_{\min}(\mu,\nu)$.

9

Making Proximities Transitive

A proximity matrix or relation on a finite universe X is a reflexive and symmetric fuzzy relation R on X. In many applications, for coherence-imposition or knowledge-learning reasons, transitivity of R with respect to a t-norm T is required. T-transitive approximation methods for proximities are especially useful in many artificial intelligence areas such as fuzzy clustering [98], non-monotonic reasoning [24], fuzzy database modelling [86] [128], decision-making and approximate reasoning [27] applications. In these cases, R must be replaced by a new relation E that also satisfies transitivity, i.e. T-indistinguishability operators. Of course, it is desirable that E be as close as possible to R. This chapter presents three reasonable -i.e. easy and rapid-ways to find close transitive relations to R when the t-norm is continuous Archimedean,as well as a fourth method for the minimum t-norm.

There are, of course, several ways to calculate the closeness of two fuzzy relations, many of them related to some metric. In this chapter, we propose a way that is related to the natural indistinguishability operator E_T associated with T, such that the degree of closeness or similarity between two fuzzy relations R and S is calculated by aggregating the similarity of their respective entries using the quasi-arithmetic mean generated by an additive generator of T.

In addition, the Euclidean metric will be used as an alternative method to compare fuzzy relations.

Trying to find the closest E to R can be very expensive. Indeed, if n is the cardinality of the universe X, the transitivity of T-indistinguishability operators can be modelled by $3 \cdot \binom{n}{3}$ inequalities, and they lie in the region of the $\binom{n}{2}$-dimensional space defined by them. The calculation of E becomes, then, a non-linear programming problem. For the Łukasiewicz t-norm, and using the Euclidean distance to compare fuzzy relations, the problem is a classical quadratic non-linear programming one, and standard methods can be used to solve it. Also, for the Product t-norm, standard non-linear programming algorithms can be used [5]. For other Archimedean t-norms or distances, in

J. Recasens: Indistinguishability Operator, STUDFUZZ 260, pp. 163–176.
springerlink.com

order to measure the similarity between fuzzy relations, simpler methods are required to find an E that is close to R.

Usually, the proximity relation R is approximated by its transitive closure $\overline{R}$, by one of its transitive openings, or by using the Representation Theorem to obtain, in this case, a T-indistinguishability operator $\underline{R}$ smaller than or equal to R. All of the values of the transitive closure are greater than or equal to the corresponding values of R, while all of the values of the transitive openings and $\underline{R}$ are smaller than or equal to the corresponding values of R. Methods for finding T-indistinguishability operators with some values greater and some values smaller than those of R will generate better approximations of R.

It appears reasonable to aggregate $\overline{R}$ and a transitive opening B or $\underline{R}$ to obtain a new T-indistinguishability operator closer to R than $\overline{R}$, B or $\underline{R}$. Section 9.1 will explore this idea.

If E is a T-indistinguishability operator, then the powers $E^{(p)}$ $p > 0$ of E with respect to the t-norm T are T-indistinguishability operators as well (see Section 9.2 for the definition of $E^{(p)}$). This allows us to increase or decrease the values of E, since $E^{(p)} \leq E^{(q)}$ for $p \geq q$. So, we can decrease the values of the transitive closure or increase the values of an operator smaller than R to find better approximations of it. Section 9.2 is devoted to this idea.

In Section 9.3, non linear programming techniques are applied to find the closest T-indistinguishability operator to a given proximity with respect to the Euclidean distance for the Product and the Łukasiewicz t-norms.

The methods used for Archimedean t-norms cannot be applied to the minimum t-norm. Section 9.4 provides an easy algorithm for computing better approximations of a fuzzy relation by a min-indistinguishability operator than its transitive closure or its transitive openings.

9.1 Aggregating the T-Transitive Closure and a T-Indistinguishability Operator Smaller Than or Equal to R

Given a proximity relation R on X, if T-transitivity is required, it is necessary to replace it by a T-indistinguishability operator E. In this case, we want to find E as close as possible to R, where the closeness or similarity between fuzzy relations can be defined in many different ways.

Let X be a finite set of cardinality n. Ordering its elements linearly, we can view the fuzzy subsets of X as vectors: $X = \{x_1, ..., x_n\}$ and a fuzzy set μ is the vector $(\mu(x_1), ..., \mu(x_n))$. A proximity relation R on X can be represented by a matrix (also called R) determined by the $\binom{n}{2}$ entries r_{ij} $1 \leq i < j \leq n$ of R above the diagonal.

Proposition 9.1. *Let $E = (e_{ij})_{i,j=1,\ldots,n}$ be a proximity matrix on a set X of cardinality n and T a continuous Archimedean t-norm with additive generator* *t. E is a T-indistinguishability operator if and only if for all i, j, k $1 \leq i < j < k \leq n$*

$$t(e_{ij}) + t(e_{jk}) \geq t(e_{ik})$$
$$t(e_{ij}) + t(e_{ik}) \geq t(e_{jk})$$
$$t(e_{ik}) + t(e_{jk}) \geq t(e_{ij}).$$

Given a proximity matrix R, we must then search for (one of) the closest matrices E to R satisfying the last $3 \cdot \binom{n}{3}$ inequalities, which is a non-linear programming problem. This is a hard problem in general though for the Łukasiewicz and the Product t-norms and the Euclidean distance well known algorithms can be applied (see Section 9.3).

In this and the next sections we propose alternative methods to obtain not the best but reasonably good approximations of proximity relations by T-indistinguishability operators when T is a continuous Archimedean t-norm.

Given a continuous monotonic map $t : [0,1] \to [-\infty, \infty]$ and p, q positive integers with $p + q = 1$, the weighted quasi-arithmetic mean $m_t^{p,q}$ generated by t and with weights p and q of $x, y \in [0,1]$ is

$$m_t^{p,q}(x,y) = t^{-1}\left(p \cdot t(x) + q \cdot t(y)\right).$$

In Chapter 8 it has been proved that if E, E' are T-indistinguishability operators for T a continuous Archimedean t-norm with additive generator t, and $m_t^{p,q}$ the weighted quasi-arithmetic mean generated by t, then $m_t^{p,q}(E, E')$ is a T-indistinguishability operator. Thanks to this result, given a proximity matrix R we can calculate its transitive closure $\overline{R}$ and a smaller T-indistinguishability operator than R, for example $\underline{R}$ obtained by its columns using the Representation Theorem and find the weights $p, 1-p$ to obtain the closest average of $\overline{R}$ and $\underline{R}$ to R.

The similarity between two fuzzy relations on X will be calculated in the following way.

Definition 9.2. *Let T be a continuous Archimedean t-norm with additive generator t and R, S two fuzzy relations on a finite set X of cardinality n. The degree $DS(R,S)$ of similarity or closeness between R and S is defined by*

$$DS(R,S) = t^{-1}\left(\frac{\sum_{1 \leq i,j \leq n} |t(r_{ij}) - t(s_{ij})|}{n^2}\right).$$

Proposition 9.3. *DS is a T-indistinguishability operator on the set of fuzzy relations on X.*

Corollary 9.4. *Let $R = (r_{ij})$ be a proximity matrix on a finite set X of cardinality n, T a continuous Archimedean t-norm with additive generator t, $\overline{R} = (\overline{r}_{i,j})$ its transitive closure, $\underline{R} = (\underline{r}_{ij})$ the T-indistinguishability operator obtained from R with the Representation Theorem, $p \in [0,1]$ and*

$m_t^{p,1-p}(\overline{R}, \underline{R})$ *the* T*-indistinguishability operator quasi-arithmetic mean of* $\overline{R}$ *and* $\underline{R}$ *with weights* p *and* $1-p$*. Then*

$$DS(R, m_t^{p,1-p}(\overline{R}, \underline{R})) =$$

$$t^{-1}\left(\frac{\sum_{1\leq i,j\leq n} \left|p \cdot t(\overline{r}_{ij}) + (1-p)\cdot t(\underline{r}_{ij}) - t(r_{ij})\right|}{n^2}\right).$$

We are looking for the value or values of p that maximize the last equality. Since t^{-1} is a decreasing map, this is equivalent to minimize

$$\sum_{1\leq i,j\leq n} \left|p \cdot t(\overline{r}_{ij}) + (1-p)\cdot t(\underline{r}_{ij}) - t(r_{ij})\right|$$

and, since R is reflexive and symmetric, is equivalent to minimize

$$f(p) = \sum_{1\leq i<j\leq n} \left|p \cdot t(\overline{r}_{ij}) + (1-p)\cdot t(\underline{r}_{ij}) - t(r_{ij})\right|$$

Since each summand of f is a concave function, so is f. Since furthermore f is piecewise linear, its set of minima consists of a single point or a closed interval.

Proposition 9.5. *The computation of the* T*-indistinguishability operator* $m_t^{p,q}(\overline{R}, \underline{R})$ *with maximum* $DS(R, m_t^{p,q}(\overline{R}, \underline{R}))$ *can be done taking* $O(n^3)$ *time complexity.*

Proof. The computation of $\overline{R}$ and $\underline{R}$ can be done in $O(n^3)$ complexity time [101].

The addition (aggregation of distances) takes $O(n^2)$ time complexity.

The minimization of $f(p)$ takes at most $O(n^2)$ time complexity.

So the most complex part of this process is the computation of $\overline{R}$ and $\underline{R}$, which still takes $O(n^3)$ complexity time.

Example 9.6. Let X be a set of cardinality 7 and R the proximity relation given by

$$R = \begin{pmatrix} 1 & 1 & 0.3 & 0.3 & 0.1 & 0.3 & 0.4 \\ 1 & 1 & 0.6 & 0.4 & 0.5 & 0.4 & 0.2 \\ 0.3 & 0.6 & 1 & 0.1 & 0.3 & 0.2 & 0.5 \\ 0.3 & 0.4 & 0.1 & 1 & 1 & 1 & 1 \\ 0.1 & 0.5 & 0.3 & 1 & 1 & 1 & 1 \\ 0.3 & 0.4 & 0.2 & 1 & 1 & 1 & 1 \\ 0.4 & 0.2 & 0.5 & 1 & 1 & 1 & 1 \end{pmatrix}.$$

Then, for T the Łukasiewicz t-norm,

$$\overline{R} = \begin{pmatrix} 1 & 1 & 0.6 & 0.4 & 0.5 & 0.4 & 0.4 \\ 1 & 1 & 0.6 & 0.5 & 0.5 & 0.5 & 0.5 \\ 0.6 & 0.6 & 1 & 0.5 & 0.5 & 0.5 & 0.5 \\ 0.4 & 0.5 & 0.5 & 1 & 1 & 1 & 1 \\ 0.5 & 0.5 & 0.5 & 1 & 1 & 1 & 1 \\ 0.4 & 0.5 & 0.5 & 1 & 1 & 1 & 1 \\ 0.4 & 0.5 & 0.5 & 1 & 1 & 1 & 1 \end{pmatrix}$$

and

$$\underline{R} = \begin{pmatrix} 1 & 0.6 & 0.3 & 0.1 & 0.1 & 0.1 & 0.1 \\ 0.6 & 1 & 0.3 & 0.2 & 0.1 & 0.2 & 0.2 \\ 0.3 & 0.3 & 1 & 0.1 & 0.1 & 0.1 & 0.1 \\ 0.1 & 0.2 & 0.1 & 1 & 0.8 & 0.9 & 0.6 \\ 0.1 & 0.1 & 0.1 & 0.8 & 1 & 0.8 & 0.7 \\ 0.1 & 0.2 & 0.1 & 0.9 & 0.8 & 1 & 0.7 \\ 0.1 & 0.2 & 0.1 & 0.6 & 0.7 & 0.7 & 1 \end{pmatrix}.$$

$$\begin{aligned} f(p) = {} & |0.4p| + |0.3p - 0.3| + |0.3p - 0.1| + |0.4p - 0.4| \\ & + |0.3p - 0.1| + |0.3p| + |0.3p| + |0.3p - 0.1| + |0.4p| + |0.3p - 0.1| \\ & + |0.3p - 0.3| + |0.4p - 0.4| + |0.4p - 0.2| + |0.4p - 0.3| + |0.4p| \\ & + |0.2p| + |0.1p| + |0.4p| + |0.2p| + |0.3p| + |0.3p| \end{aligned}$$

which attains its minimum for $p = \frac{1}{3}$.

A good T-transitive approximation of R (for T the Łukasiewicz t-norm) is then

$$m_t^{\frac{1}{3},\frac{2}{3}}(\overline{R}, \underline{R}) = \begin{pmatrix} 1 & 0.733 & 0.4 & 0.2 & 0.233 & 0.2 & 0.2 \\ 0.733 & 1 & 0.4 & 0.3 & 0.233 & 0.3 & 0.3 \\ 0.4 & 0.4 & 1 & 0.233 & 0.233 & 0.233 & 0.233 \\ 0.2 & 0.3 & 0.233 & 1 & 0.867 & 0.933 & 0.733 \\ 0.233 & 0.233 & 0.233 & 0.867 & 1 & 0.867 & 0.8 \\ 0.2 & 0.3 & 0.233 & 0.933 & 0.867 & 1 & 0.8 \\ 0.2 & 0.3 & 0.233 & 0.733 & 0.8 & 0.8 & 1 \end{pmatrix}.$$

The degree of closeness between two fuzzy relations can also be calculated using the Euclidean distance.

Definition 9.7. *Let $R = (r_{ij})$ and $S = (s_{ij})$ be two fuzzy relations on a finite set X of cardinality n. The Euclidean distance D between R and S is*

$$D(R, S) = \left(\sum_{1 \leq i,j \leq n} (r_{ij} - s_{ij})^2 \right)^{\frac{1}{2}}$$

Proposition 9.8. *Let $R = (r_{ij})$ be a proximity matrix on a finite set X of cardinality n, T a continuous Archimedean t-norm with additive* *generator t, $\overline{R} = (\overline{r}_{i,j})$ its transitive closure, $\underline{R} = (\underline{r}_{ij})$ the T-indistinguishability operator obtained from R with the Representation Theorem, $p \in [0,1]$ and $m_t^{p,1-p}(\overline{R},\underline{R})$ the T-indistinguishability operator quasi-arithmetic mean of $\overline{R}$ and $\underline{R}$ with weights p and $1-p$. Then*

$$D(R, m_t^{p,1-p}(\overline{R},\underline{R})) =$$

$$\left(\sum_{1\leq i,j\leq n} \left(t^{-1}\left(p\cdot t\left(\overline{r}_{ij}\right) + (1-p)\cdot t\left(\underline{r}_{ij}\right)\right) - t(r_{ij})\right)^2 \right)^{\frac{1}{2}}.$$

Proposition 9.9. *Let T be the Łukasiewicz t-norm and R a proximity on a set X of cardinality n. The closest $m_t^{p,1-p}(\overline{R},\underline{R})$ to R is attained for*

$$p = \frac{\sum_{1\leq i<j\leq n}\left(\overline{r}_{ij}-\underline{r}_{ij}\right)\left(r_{ij}-\underline{r}_{ij}\right)}{\sum_{1\leq i<j\leq n}\left(\overline{r}_{ij}-\underline{r}_{ij}\right)^2}$$

Proof. Due to symmetry and reflexivity, it is enough to minimize

$$f(p) = \sum_{1\leq i<j\leq n}\left(p\left(\overline{r}_{ij}-\underline{r}_{ij}\right)+\underline{r}_{ij}-r_{ij}\right)^2.$$

$$f'(p) = 2\sum_{1\leq i<j\leq n}\left(p\left(\overline{r}_{ij}-\underline{r}_{ij}\right)+\underline{r}_{ij}-r_{ij}\right)\left(\underline{r}_{ij}-r_{ij}\right) = 0$$

and

$$p = \frac{\sum_{1\leq i<j\leq n}\left(\overline{r}_{ij}-\underline{r}_{ij}\right)\left(r_{ij}-\underline{r}_{ij}\right)}{\sum_{1\leq i<j\leq n}\left(\overline{r}_{ij}-\underline{r}_{ij}\right)^2}.$$

Example 9.10. Let X be a set of cardinality 4 and R the proximity relation on X given by

$$R = \begin{pmatrix} 1 & 0.8 & 0.2 & 0.4 \\ 0.8 & 1 & 0.7 & 0.1 \\ 0.2 & 0.7 & 1 & 0.6 \\ 0.4 & 0.1 & 0.6 & 1 \end{pmatrix}.$$

If T is the Łukasiewicz t-norm, the closest T-indistinguishability operator of the type $m_t^{p,1-p}(\overline{R},\underline{R})$ (with respect to the Euclidean distance) is attained for $p = 0.6388889$.

A good T-approximation of R is then

$$\begin{pmatrix} 1 & 0.6917 & 0.3917 & 0.3639 \\ 0.6917 & 1 & 0.5917 & 0.2278 \\ 0.3917 & 0.5917 & 1 & 0.5278 \\ 0.3639 & 0.2278 & 0.5278 & 1 \end{pmatrix}.$$

9.2 Applying a Homotecy to a T-Indistinguishability Operator

In this section, the fact that the power of a T-indistinguishability operator in the sense of Definition 9.13 is again a T-indistinguishability operator will be exploited to modify the entries of $\overline{R}$ or $\underline{R}$ to find a better approximation of R.

Definition 9.11. *Given a t-norm T, $T(\overbrace{x, x, ...x}^{n\ times})$ -the n-th power of x- will be denoted by $x_T^{(n)}$ or simply by $x^{(n)}$ if the t-norm is clear.*

The n-th root $x_T^{(\frac{1}{n})}$ of x with respect to T is defined by

$$x_T^{(\frac{1}{n})} = \sup\{z \in [0,1] \mid z_T^{(n)} \leq x\}$$

and for $m, n \in \mathbb{N}$, $x_T^{(\frac{m}{n})} = \left(x_T^{(\frac{1}{n})}\right)_T^{(m)}$.

Lemma 9.12. [83] *If $k, m, n \in \mathbb{N},\ k, n \neq 0$ then $x_T^{(\frac{km}{kn})} = x_T^{(\frac{m}{n})}$.*

The powers $x_T^{(\frac{m}{n})}$ can be extended to irrational exponents in a straightforward way.

Definition 9.13. *If $r \in \mathbb{R}^+$ is a positive real number, let $\{a_n\}_{n\in\mathbb{N}}$ be a sequence of rational numbers with $\lim_{n\to\infty} a_n = r$. For any $x \in [0,1]$, the power $x_T^{(r)}$ is*

$$x_T^{(r)} = \lim_{n\to\infty} x_T^{(a_n)}.$$

Continuity assures the existence of the last limit and independence of the sequence $\{a_n\}_{n\in\mathbb{N}}$.

Proposition 9.14. *Let T be a continuous Archimedean t-norm with additive generator t, $x \in [0,1]$ and $r \in \mathbb{R}^+$. Then*

$$x_T^{(r)} = t^{[-1]}(rt(x)).$$

Proof. Due to the continuity of t we need to prove it only for rational r.

If r is a natural number m, then trivially $x_T^{(m)} = t^{[-1]}(mt(x))$.

If $r = \frac{1}{n}$ with $n \in \mathbb{N}$, then $x_T^{(\frac{1}{n})} = z$ with $z_T^{(n)} = x$ or $t^{[-1]}(nt(z)) = x$ and $x_T^{(\frac{1}{n})} = t^{[-1]}\left(\frac{t(x)}{n}\right)$.

For a rational number $\frac{m}{n}$,

$$x_T^{(\frac{m}{n})} = \left(x_T^{(\frac{1}{n})}\right)_T^{(m)} = t^{[-1]}\left(\frac{m}{n}t(x)\right).$$

Proposition 9.15. *Let T be a continuous t-norm, E a T-indistinguishability operator on X and $p > 0$. Then $E^{(p)}$ is a T-indistinguishability operator.*

Proof

- If $n \in \mathbb{N}$, then $E^{(n)}$ is a T-indistinguishability operator as a consequence of Proposition 8.1.
- If $n \in \mathbb{N}$, then $E^{(\frac{1}{n})}$ is a T-indistinguishability operator:
  Reflexivity ans symmetry are trivial.
  Transitivity: If $E^{(\frac{1}{n})} = F$, then $F^{(n)} = E$. Since E is a T-indistinguishability operator, $\forall x, y, z \in X$

$$F^{(n)}(x,z) \leq T(F^{(n)}(x,y), F^{(n)}(y,z)) = (T(F(x,y), F(y,z)))^{(n)}.$$

$$(F^{(n)}(x,z))^{(\frac{1}{n})} \leq ((T(F(x,y), F(y,z)))^{(n)})^{(\frac{1}{n})}$$

  and from Lemma 9.12

$$F(x,z) \leq T(F(x,y), F(y,z)).$$

- If $m, n \in \mathbb{N}$, then $E^{(\frac{m}{n})}$ is a T-indistinguishability operator: Indeed, $E^{(\frac{m}{n})} = (E^{(\frac{1}{n})})^{(m)}$.
- Continuity assures the result for any positive real number p.

We will say that $E^{(p)}$ is obtained from E by a homotecy of power p.

Example 9.16

- If T is a continuous Archimedean t-norm with additive generator t and E a T-indistinguishability operator, then $t^{[-1]}(p \cdot t(E))$ is a T-indistinguishability operator.
- If T is the Łukasiewicz t-norm and E a T-indistinguishability operator, then $\max(0, 1 - p + p \cdot E)$ is a T-indistinguishability operator.
- If T is the Product t-norm and E a T-indistinguishability operator, then E^p is a T-indistinguishability operator.

Let $R = (r_{ij})$ be a proximity matrix on a set X of cardinality X, $p > 0$ and $E = (e_{ij})$ a T-indistinguishability operator on X with T a continuous Archimedean t-norm with additive generator t. Then

$$DS(R, E^{(p)}) = t^{-1}\left(\frac{\sum_{1 \leq i,j \leq n} |t(r_{ij}) - p \cdot t(e_{ij}))|}{n^2}\right).$$

To maximize the previous expression is equivalent to minimize

$$\sum_{1 \leq i,j \leq n} |t(r_{ij}) - p \cdot t(e_{ij}))|.$$

Since R is reflexive and symmetric, this is equivalent to minimize

$$g(p) = \sum_{1\leq i<j\leq n} |t(r_{ij}) - p \cdot t(e_{ij}))| \,.$$

g is a sum of concave functions in $[0,1]$ and therefore has one minimum or a close interval of minima.

Example 9.17. Let us consider the same matrix of Example 9.10.

$$R = \begin{pmatrix} 1 & 0.8 & 0.2 & 0.4 \\ 0.8 & 1 & 0.7 & 0.1 \\ 0.2 & 0.7 & 1 & 0.6 \\ 0.4 & 0.1 & 0.6 & 1 \end{pmatrix}.$$

Then, for T the Łukasiewicz t-norm,

$$\underline{R} = \begin{pmatrix} 1 & 0.5 & 0.2 & 0.3 \\ 0.5 & 1 & 0.4 & 0.1 \\ 0.2 & 0.4 & 1 & 0.4 \\ 0.3 & 0.1 & 0.4 & 1 \end{pmatrix}.$$

$$g(p) = |0.5 \cdot p - 0.2| + |0.8 \cdot p - 0.8| + |0.7 \cdot p - 0.6| + \\ |0.6 \cdot p - 0.3| + |0.9 \cdot p - 0.9| + |0.6 \cdot p - 0.4|$$

which attains its minimum for $p = 0.857$.

A good approximation of R is then

$$\underline{R}^{(0.857)} = \begin{pmatrix} 1 & 0.5715 & 0.3144 & 0.4001 \\ 0.5715 & 1 & 0.4858 & 0.2287 \\ 0.3144 & 0.4858 & 1 & 0.4858 \\ 0.4001 & 0.2287 & 0.4858 & 1 \end{pmatrix}.$$

If we consider the Euclidean distance between R and the power $E^{(p)}$ of a T-indistinguishability operator $E = (e_{ij})$, then we can prove the following result.

Proposition 9.18

$$D(R, E^{(p)}) = \left(\sum_{1\leq i,j\leq n} \left(t^{-1}\left(p \cdot t\left(e_{ij}\right)\right) - r_{ij}\right)^2 \right)^{\frac{1}{2}}.$$

Example 9.19 Continuing the last example, $D(R, \overline{R}^{(p)})$ attains its maximum for $p = 1.208633$ and $D(R, \underline{R}^{(p)})$ for $p = 0.821306$.

Good approximations of R are therefore

$$\overline{R}^{(1.208633)} = \begin{pmatrix} 1 & 0.7583 & 0.3957 & 0.2748 \\ 0.7583 & 1 & 0.6374 & 0.1540 \\ 0.3957 & 0.6374 & 1 & 0.5165 \\ 0.2748 & 0.1540 & 0.5165 & 1 \end{pmatrix}.$$

and

$$\underline{R}^{(0.821306)} = \begin{pmatrix} 1 & 0.8357 & 0.5893 & 0.5072 \\ 0.8357 & 1 & 0.7536 & 0.4251 \\ 0.5893 & 0.7536 & 1 & 0.6715 \\ 0.5072 & 0.4251 & 0.6715 & 1 \end{pmatrix}.$$

9.3 Using Non-linear Programming Techniques

As it was shown previously, trying to find the best approximation to a given proximity relation becomes a non-linear programming problem.

The inequalities of Proposition 9.1 are especially simple for the Łukasiewicz and the Product t-norms. This will allow us to find the best approximation of a given proximity for these two t-norms with respect to the Euclidean distance.

Proposition 9.20. *Let $E = (e_{ij})_{i,j=1,\ldots,n}$ be a proximity matrix on a set X of cardinality n. E is a T-indistinguishability operator where T is the Łukasiewicz t-norm if and only if for all i, j, k $1 \leq i < j < k \leq n$*

$$\begin{aligned} e_{ij} + e_{jk} - e_{ik} &\leq 1 \\ e_{ij} + e_{ik} - e_{jk} &\leq 1 \\ e_{ik} + e_{jk} - e_{ij} &\leq 1. \end{aligned}$$

Proposition 9.21. *Let $E = (e_{ij})_{i,j=1,\ldots,n}$ be a proximity matrix on a set X of cardinality n. E is a T-indistinguishability operator where T is the Product t-norm if and only if for all i, j, k $1 \leq i < j < k \leq n$*

$$\begin{aligned} e_{ij} \cdot e_{jk} &\leq e_{ik} \\ e_{ij} \cdot e_{ik} &\leq e_{jk} \\ e_{ik} \cdot e_{jk} &\leq e_{ij}. \end{aligned}$$

Let $R = (r_{ij})_{i,j=1,\ldots,n}$ be a proximity relation on a set X of cardinality n. If we want to find the best approximation of R by a T-indistinguishability operator $E = (e_{ij})_{i,j=1,\ldots,n}$ with respect to the Euclidean distance we must minimize

$$D(R, E) = \sqrt{\sum_{i<j} (e_{ij} - r_{ij})^2}.$$

This is equivalent to minimize

$$f(R,E)=\sum_{i<j}(e_{ij}-r_{ij})^2$$

where e_{ij} $i,j=1,...,n$ are subject to the conditions of Proposition 9.1.

Proposition 9.22. *Let $R=(r_{ij})_{i,j=1,\ldots,n}$ be a proximity relation on a set X of cardinality n. The closest T-indistinguishability operator to R with respect to the Euclidean distance when T is the Łukasiewicz t-norm is the solution of minimizing*

$$f(R,E)=\sum_{i<j}(e_{ij}-r_{ij})^2$$

where e_{ij} $i,j=1,...,n$ are subject to the linear inequalities of Proposition 9.20.

This is a standard quadratic linear problem and there are some algorithms to solve it [5].

Example 9.23. Let us consider the matrix

$$R=\begin{pmatrix}1 & 0.8 & 0.2 & 0.4\\ 0.8 & 1 & 0.7 & 0.1\\ 0.2 & 0.7 & 1 & 0.6\\ 0.4 & 0.1 & 0.6 & 1\end{pmatrix}$$

of Example 9.10. The closest $E=(e_{ij})_{i,j=1,\ldots,4}$ must minimize

$$f(R,E)=(e_{ij}-0.8)^2+(e_{ij}-0.2)^2+(e_{ij}-0.4)^2+$$

$$(e_{ij}-0.7)^2+(e_{ij}-0.1)^2+(e_{ij}-0.6)^2$$

subject to the conditions of Proposition 9.20.

The matrix E is

$$E=\begin{pmatrix}1 & 0.65 & 0.225 & 0.4\\ 0.65 & 1 & 0.575 & 0.225\\ 0.225 & 0.575 & 1 & 0.575\\ 0.4 & 0.225 & 0.575 & 1\end{pmatrix}.$$

The distance between E and R is 0.234521, the distance between R and $m_t(\underline{R},\overline{R})$ with $p=0.638889$ (Example 9.10) is 0.288194, the distance between R and the best homotecy of its transitive closure is 0.263665 and the distance between R and the best homotecy of the relation obtained by the Representation Theorem is 0.527253 (Example 9.19).

A similar result to Proposition 9.22 can be obtained for the Product t-norm.

Proposition 9.24. *Let $R=(r_{ij})_{i,j=1,\ldots,n}$ be a proximity relation on a set X of cardinality n. The closest T-indistinguishability operator to R with respect* *to the Euclidean distance when T is the Product t-norm is the solution of minimizing*

$$f(R,E)=\sum_{i<j}(e_{ij}-r_{ij})^2$$

where e_{ij} $i,j=1,...,n$ are subject to the inequalities of Proposition 9.21.

9.4 The Minimum t-Norm

In this section we present a simple algorithm to find a min-indistinguishability operator close to a given proximity relation which is closer than its transitive closure or any of its transitive openings.

Let us recall from Chapter 5 that if E is a min-indistinguishability operator on a finite universe X of cardinality n, then the number of different entries of the matrix representing E is smaller than or equal to n.

The algorithm to approximate proximities by min-indistinguishability operators is based on the following result.

Proposition 9.25. *Let E be a* min-*indistinguishability operator on a finite universe X of cardinality n and $a_1<a_2<...<a_k=1$ $(k\leq n)$ the entries of E. If we replace the entries by $a'_1\leq a'_2\leq ...\leq a'_k=1$ respectively, we obtain a new* min-*indistinguishability operator on X.*

Proof. Reflexivity and symmetry are trivial.

Transitivity: if $\min(a_i,a_j)\leq a_k$, then $\min(a'_i,a'_j)\leq a'_k$.

The idea for finding a similarity relation close to a given proximity R is then very easy. We can calculate the min-transitive closure $\overline{R}$ of R and then we can modify the entries of $\overline{R}$ in order to minimize some distance to R or to maximize some similarity measure to R. Of course, we can calculate a transitive opening or the similarity relation obtained from R by the Representation Theorem instead of the transitive closure.

Nevertheless, the procedure is not straightforward as it is shown in the following two examples.

Example 9.26. Let us consider the proximity with matrix

$$R=\begin{pmatrix}1 & 0.7 & 0.3 & 1\\ 0.7 & 1 & 0.4 & 0.7\\ 0.3 & 0.4 & 1 & 0.8\\ 1 & 0.7 & 0.8 & 1\end{pmatrix}.$$

Its min-transitive closure is

$$\overline{R}=\begin{pmatrix}1 & 0.7 & 0.8 & 1\\ 0.7 & 1 & 0.7 & 0.7\\ 0.8 & 0.7 & 1 & 0.8\\ 1 & 0.7 & 0.8 & 1\end{pmatrix}.$$

If we replace the entries with values 0.7 and 0.8 of $\overline{R}$ by a and b respectively in order to minimize the Euclidean distance D between R and the new matrix, we must minimize

$$f(a,b) = (a-0.7)^2 + (a-0.4)^2 + (a-0.7)^2 + (b-0.8)^2 + (b-0.3)^2.$$

$$\frac{\partial f}{\partial a} = 2(a-0.7) + 2(a-0.4) + 2(a-0.7) = 0$$
$$\frac{\partial f}{\partial b} = 2(b-0.8) + 2(b-0.3) = 0$$

and

$$a = \frac{0.7+0.7+0.4}{3} = 0.6$$
$$b = \frac{0.8+0.3}{2} = 0.55$$

obtaining

$$\begin{pmatrix} 1 & 0.6 & 0.55 & 1 \\ 0.6 & 1 & 0.6 & 0.6 \\ 0.55 & 0.6 & 1 & 0.55 \\ 1 & 0.6 & 0.55 & 1 \end{pmatrix}$$

which is not min-transitive and therefore not a min-indistinguishability operator.

The same problem may occur with approximations from below.

A possible solution in these cases is replacing the "wrongly ordered" entries by a unified value. For instance, in Example 9.26 the entries 0.6 and 0.55 can be replaced by $\frac{3\cdot 0.6+2\cdot 0.55}{5} = 0.58$. Note that, thanks to the next lemma, among all possible values 0.58 is the one who minimizes the distance between the obtained matrix and R.

Lemma 9.27. *Let $P = (x_1, x_2, ..., x_n) \in \mathbb{R}^n$. The closest point Q to P with respect to the Euclidean distance of the form $(a, a, ..., a)$ satisfies $a = \frac{x_1+x_2+...+x_n}{n}$.*

Proof. We want to minimize $d(P,Q) =$

$$\sqrt{(a-x_1)^2 + (a-x_2)^2 + ... + (a-x_n)^2}$$

which is equivalent to minimize

$$f(a) = (a-x_1)^2 + (a-x_2)^2 + ... + (a-x_n)^2.$$

The solution of

$$f'(a) = 2(a - x_1) + 2(a - x_2) + ... + 2(a - x_n) = 0$$

is

$$a = \frac{x_1 + x_2 + ... + x_n}{n}.$$

The **algorithm** to find a close min-indistinguishability operator to a given proximity relation R on a universe of cardinality n goes then as follows.

1. Calculate the min-transitivity closure $\overline{R}$ or a lower approximation of R.
2. Order the entries $a_1 < a_2 < ... < a_k = 1$ $(k \leq n)$ of $\overline{R}$.
3. Replace every a_i by the weighted arithmetic mean a'_i of the entries of R that are in the same place than a_i.
4. If $a'_1 \leq a'_2 \leq ... \leq a'_k = 1$, then the desired similarity relation E is obtained by replacing the entries $a_1, a_2, ..., a_k$ of $\overline{R}$ by $a'_1 \leq a'_2 \leq ... \leq a'_k$ respectively.
5. Else, for every maximal chain $C = \{a'_i, a'_{i+1}, ..., a'_{i+j}\}$ with $a'_i > a'_{i+j}$, replace all the elements of C by the weighted mean a_C of them, weighting every a'_l of C by the number of entries of R that correspond a'_l. Replacing the elements of C by a_C in E the desired min-indistinguishability operator is obtained.

10

Fuzzy Functions

Fuzzy functions fuzzify the concept of a function between two universes. They have been used in various fields, including vague algebras [37], fuzzy numbers [71], vague lattices and quantum mechanics, and have proven useful to the understanding of approximate reasoning [38], the analysis of input/output systems, fuzzy interpolation [38], [14] and reasoning based on fuzzy rules [80].

When fuzzy functions are used, it is assumed that the uncertainty (or fuzziness) of the domain and the image appears or is modelled by a fuzzy equivalence relation or indistinguishability operator. The presence of an indistinguishability operator on a universe determines its granules, and fuzzy functions are compatible with the granularity of the domain and the image in the sense that, roughly speaking, they map granules to granules.

Extensional crisp functions generate fuzzy functions in a very natural way and, reciprocally, crisp functions can be generated from a fuzzy function. This aspect will be explored after some definitions concerning fuzzy functions and extensionality are presented.

We will prove the existence of maximal fuzzy functions and demonstrate that every fuzzy function is contained in (at least) a maximal one. The interest of the result lies in the fact that maximal fuzzy functions deal with large granules and manage the largest amount of uncertainty, which justifies their prudent use in some situations.

In the 1970s, researchers studied a very interesting kind of fuzzy function, in which the domain did not present uncertainty. This means that crisp equality is the indistinguishability operator of the domain. These functions will be revisited within the framework of the general definition of fuzzy function.

10.1 Fuzzy Functions

In this section the definition of fuzzy function and extensionality will be given and some consequences of them are explained.

J. Recasens: Indistinguishability Operator, STUDFUZZ 260, pp. 177–188.
springerlink.com

A fuzzy function between X and Y is a fuzzy relation $R : X \times Y \to [0,1]$ that is compatible with the granularity on these sets generated by T-indistinguishability operators defined on them. $R(x,y)$ is interpreted as the degree in which y is the image of x.

Definition 10.1. *Let E and F be two T-indistinguishability operators on X and Y, respectively. A fuzzy relation $R \in [0,1]^{X\times Y}$ is called extensional with respect to E and F if and only if*

$$T(R(x,y), E(x,x'), F(y,y')) \leq R(x',y')$$

is satisfied for all $x, x' \in X$, and for all $y, y' \in Y$.

Definition 10.2. *Let E and F be two T-indistinguishability operators on X and Y respectively and let a fuzzy relation $R \in [0,1]^{X\times Y}$ be extensional with respect to E and F. Then*

1. *R is called a partial fuzzy function or partial fuzzy map if and only if*

$$T(R(x,y), R(x,y')) \leq F(y,y')$$

   is satisfied for all $x \in X$ and for all $y, y' \in Y$.
2. *A fuzzy relation $R \in [0,1]^{X\times Y}$ is said to be fully defined if and only if it fulfills the condition*

$$\bigvee_{y\in Y} R(x,y) = 1 \text{ for all } x \in X.$$

3. *A fully defined partial fuzzy function is called a fuzzy function.*
4. *A partial fuzzy function R is called a perfect fuzzy function with respect to E and F if and only if it satisfies the condition*

$$\forall\, x \in X \exists y \in Y \text{ such that } R(x,y) = 1.$$

Definition 10.3. *Let E and F be two T-indistinguishability operators on X and Y respectively. A fuzzy relation $R \in [0,1]^{X\times Y}$ is called a strong fuzzy function with respect to E and F if and only if R satisfies the condition 10.2.4 and the condition*

$$T(R(x,y), R(x',y'), E(x,x')) \leq F(y,y')$$

for all $x, x' \in X$ and for all $y, y' \in Y$.

Extensionality of a crisp function $f : X \to Y$ with respect to T-indistinguishability operators defined on X and Y is a central property for f. It is in fact a Lipschitzian property of f.

Definition 10.4. *Let E and F be two T-indistinguishability operators on X and Y respectively and $f : X \to Y$ a crisp function. f is called extensional with respect to E and F if and only if*

$$E(x, x') \leq F(f(x), f(x'))$$

holds for all $x, x' \in X$.(cf. Definition 3.78).

An extensional function f generates a fuzzy function R_f in a very natural way and reciprocally every fuzzy function R generates at least an extensional function f_R.

Proposition 10.5. *Let E and F be two T-indistinguishability operators on X and Y respectively.*

a) For a given (crisp) function $f : X \to Y$ extensional with respect to E and F a fuzzy relation $R \in [0,1]^{X \times Y}$ which satisfies the condition

$$R(x, f(x)) = 1 \text{ and } R(x, y) \leq F(f(x), y) \tag{1}$$

for all $x \in X$, $y \in Y$ is a strong fuzzy function with respect to E and F.
b) Conversely, for a given strong fuzzy function $R \in [0,1]^{X \times Y}$ with respect to E and F there exists an ordinary function $f : X \to Y$ extensional with respect to E and F satisfying the condition (1).

Proof

a) It is clear that $R(x, y)$ satisfies 10.2.4.
Using the extensionality of f and taking (1) into account,

$$\begin{aligned} &T(R(x,y), R(x',y'), E(x,x')) \\ &\leq T(F(f(x),y), f(f(x'),y'), F(f(x), f(x'))) \\ &\leq T(F(f(x),y), F(f(x),y')) \leq F(y,y') \end{aligned}$$

for all $x, x' \in X$ and $y, y' \in Y$. Therefore R is a strong fuzzy function.
b) For a given strong fuzzy function $R \in [0,1]^{X \times Y}$ with respect to E and F let us consider the subset $H = \{(x, y) \mid R(x, y) = 1\}$ of $X \times Y$. For each $x \in X$, let $H_x = \{y \in Y \mid (x, y) \in H\}$. By 10.2.4, for each $x \in X$, $\exists y \in Y$ such that $(x, y) \in H$, i.e. $H_x \neq \emptyset$. Therefore, by the axiom of choice, the family $\{H_x \mid x \in X\}$ of all subsets H_x of Y has a choice set C, i.e. C has one and only one element from each subset H_x of Y. Then, if we define the function $f : X \to Y$ such that $f(x) \in H_x \cap C$ for each $x \in X$, it is obvious that $f : X \to Y$ is well-defined. By the definition of f, since $R(x, f(x)) = 1 \ \forall x \in X$, and using the fact that R is a strong fuzzy function, we see that

$$\begin{aligned} E(x,x') &= T(R(x,f(x)), R(x',f(x')), E(x,x')) \\ &\leq F(f(x), f(x')) \ \forall x, x' \in X, \end{aligned}$$

i.e. f is an extensional function with respect to E and F. Then, exploiting the fact that $R(x, f(x)) = 1 \; \forall x \in X$, we also have

$$R(x,y) = T(R(x,y), R(x,f(x)), E(x,x)) \\ \leq F(f(x),y), \forall x \in X, \forall y \in Y.$$

Proposition 10.6. *A perfect fuzzy function is a strong fuzzy function.*

Proof

$$T(R(x,y), R(x',y')), E(x,y)) \leq T(R(x',y), R(x',y')) \leq F(y,y').$$

Proposition 10.7. *Let E and F be two T-indistinguishability operators on X and Y respectively.*

a) For a given ordinary function $f : X \to Y$ extensional with respect to E and F a fuzzy relation $R \in [0,1]^{X \times Y}$ defined by the formula

$$R(x,y) = F(f(x),y), \forall x \in X \; \forall y \in Y \tag{2}$$

is a perfect fuzzy function with respect to E and F.

b) Conversely, for a given perfect fuzzy function $R \in [0,1]^{X \times Y}$ with respect to E and F there exists an ordinary function $f : X \to Y$ extensional with respect to E and F fulfilling the equality (2).

Proof. a) R clearly verifies 10.2.4, since $R(x, f(x)) = F(f(x), f(x)) = 1$. Thanks to Proposition 10.5, R is a strong fuzzy function.
Let us show that R is extensional with respect E and F. From the extensionality of f and (2) for all $x, x' \in X$ and $y, y' \in Y$ we obtain

$$T(R(x,y), E(x,x'), F(y,y')) \leq T(F(f(x),y), F(f(x'), f(x), F(y,y')) \\ \leq F(f(x'),y') = R(x',y').$$

b) From Proposition 10.6, R is a strong fuzzy function and hence from Proposition 10.5 there exists an extensional function $f : X \to Y$ extensional with respect to E and F satisfying (1). Using this property and the extensionality of R, we obtain

$$F(f(x),y) = T(R(x,f(x)), F(f(x),y)) \leq R(x,y).$$

Therefore $R(x,y) = F(f(x),y)$.

In Chapter 11 we will need to deal with partial crisp functions, i.e. with functions $f : X \to Y$ whose domain (dom(f)) is a subset of X. In this case, the fuzzy function generated by f is slightly more complicated.

Proposition 10.8. *Let E and F be T-indistinguishability operators on X and Y respectively and $f : X \to Y$ a partial function with domain* dom(f). *The relation R_f defined by*

$$R_f(x,y) = \sup_{x' \in \text{ dom}(f)} T(E(x,x'), F(f(x'),y))$$

is a partial fuzzy function.

Proof. Let us first prove that R_f is extensional.

$$\begin{aligned} T(R_f(x,y), E(x,x')) &= \sup_{x'' \in \text{ dom}(f)} T(E(x,x''), E(x,x'), F(y, f(x''))) \\ &\leq \sup_{x'' \in \text{ dom}(f)} T(E(x',x''), F(y, f(x''))) \\ &= R_f(x',y). \end{aligned}$$

$$\begin{aligned} T(R_f(x,y), F(y,y')) &= \sup_{x' \in \text{ dom}(f)} T(E(x,x'), F(y, f(x'')), F(y,y')) \\ &\leq \sup_{x' \in \text{ dom}(f)} T(E(x,x'), F(y', f(x'))) \\ &= R_f(x,y'). \end{aligned}$$

Let us now prove 10.2.1.

$$\begin{aligned} &T(R_f(x,y), R_f(x,y')) \\ &= \sup_{x',x'' \in \text{ dom}(f)} T(E(x,x'), F(y, f(x')), E(x,x''), F(y, f(x''))) \\ &\leq \sup_{x',x'' \in \text{ dom}(f)} T(E(x',x''), F(y, f(x')), F(y, f(x''))) \\ &\leq \sup_{x',x'' \in \text{ dom}(f)} T(F(f(x'), f(x'')), F(y, f(x'), F(y, f(x''))) \\ &\leq F(y,y'). \end{aligned}$$

Note that if the domain of f is X, then we recover the previous fuzzy function $R_f(x,y) = F(y, f(x))$. Indeed, in this case,

$$\begin{aligned} R_f(x,y) &= \sup_{x' \in X} T(E(x,x'), F(y, f(x'))) \\ &\leq \sup_{x' \in X} T(F(f(x), f(x')), F(y, f(x'))) \\ &\leq F(f(x), y). \end{aligned}$$

But for $x' = x$ we obtain the equality.

Definition 10.9. *Let E and F be two T-indistinguishability operators on X and Y respectively.*

1. *For a given function $f : X \to Y$ extensional with respect to E and F the perfect fuzzy function $R \in [0,1]^{X \times Y}$ with respect to E and F given by the formula $R_f(x,y) = F(f(x),y))$ is called an $E - F$ vague description of f, and it is denoted by* $\text{vag}(f)$.

2. For a given perfect (strong) fuzzy function $R \in [0,1]^{X \times Y}$ with respect to E and F, an ordinary function $f : X \to Y$ extensional with respect to E and F satisfying $R_f(x,y) = F(f(x),y))$ $(R(x,f(x)) = 1,\ R_f(x,y) \leq F(f(x),y)))$ is called an ordinary description of R.

The set of all ordinary descriptions of R is denoted by $ORD(R)$.

$ORD(R)$ has more than one element in general but if F separates points, then there is only a function f in $ORD(R)$. In this case we will simply denote $f = \mathrm{ord}(R)$.

Proposition 10.10. *Let E and F be two T-indistinguishability operators on X and Y respectively, $f : X \to Y$ an extensional with respect to E and F, $R \in [0,1]^{X \times Y}$ a perfect fuzzy function with respect to E and F and $S \in [0,1]^{X \times Y}$ a strong fuzzy function with respect to E and F. The following properties are satisfied:*

1. $S \leq \bigwedge_{g \in ORD(S)} \mathrm{vag}(g)$
2. $R = \mathrm{vag}(g),\ \forall g \in ORD(R)$
3. $f \in ORD(\mathrm{vag}(f))$
4. *If F separates points, then*
   a. $S \leq \mathrm{vag}(\mathrm{ord}(S))$
   b. $R = \mathrm{vag}(\mathrm{ord}(R))$
   c. $f = \mathrm{ord}(\mathrm{vag}(f))$.

Proof. It follows directly from the definitions of $\mathrm{vag}(f)$ and $\mathrm{ord}(R)$, $\mathrm{ord}(S)$.

For the sake of simplicity, until the end of this chapter we will assume that all T-indistinguishability operators separate points.

Going back to Definition 10.1, it simply states that R (as a fuzzy subset of $X \times Y$) is extensional with respect to the T-indistinguishability $E \times F$ on $X \times Y$ defined by

$$(E \times F)\left((x,y),(x',y')\right) = T\left(E(x,x'),F(y,y')\right).$$

It can be split into two parts:

a) $T(E(x,x'),R(x,y)) \leq R(x',y)$
b) $T(F(y,y'),R(x,y)) \leq R(x,y')$.

a) expresses that fixing $y_0 \in Y$ the fuzzy subset $R(\cdot,y_0)$ of X is extensional with respect to E while b) states that fixing $x_0 \in X$, the fuzzy subset $R(x_0,\cdot)$ of Y is extensional with respect to F.

The next proposition translates a) when X and Y are the unit interval and the T-indistinguishability operators powers of the natural one with T a continuous Archimedean t-norm.

Proposition 10.11. *Let T be a continuous Archimedean t-norm with additive generator t and R a fuzzy function on $[0,1] \times [0,1]$, the first interval endowed with the T-indistinguishability operator $E_T^{(k)}$ and the second one with $E_T^{(l)}$.*

a) For any $y_0 \in Y$ $t \circ R(\cdot, y_0)$ is an $(E_T^{(k)}, E_T)$-extensional function.
b) For any $x_0 \in X$ $t \circ R(x_0, \cdot)$ is an $(E_T^{(l)}, E_T)$-extensional function.

Proof. a)

$$t^{-1}(t(t^{-1}(k\,|t(x) - t(x')|) + t(R(x, y_0))) \leq R(x', y_0)$$

and

$$k\,|t(x) - t(x')| \geq t(R(x', y_0)) - t(R(x, y_0)).$$

Similarly,

$$k\,|t(x) - t(x')| \geq t(R(x, y_0)) - t(R(x', y_0))$$

and hence

$$k\,|x - x'| \geq |t(R(x', y_0)) - t(R(x, y_0))|\,.$$

Therefore,

$$t^{-1}\left(k\,|x - x'|\right) \leq t^{-1}\left(|t(R(x', y_0)) - t(R(x, y_0))|\right)$$

and

$$E_T^{(k)}(x, x') \leq E_T(R(x', y_0), R(x, y_0)).$$

b) is similar to a).

10.2 Composition of Fuzzy Functions

In this section the composition of fuzzy functions and some properties related to injectivity, exhaustivity and bijectivity will be studied.

Definition 10.12. *Let E, F, G be T-indistinguishability operators on X, Y and Z respectively and $R \in [0,1]^{X \times Y}$ and $S \in [0,1]^{Y \times Z}$ fuzzy relations. The composition $S \circ R$ is the fuzzy relation on $X \times Z$ defined for all $x \in X$ and $z \in Z$ by*

$$(S \circ R)(x, z) = \sup_{y \in Y} T(R(x, y), S(y, z)).$$

Proposition 10.13. *If R and S are perfect (strong) fuzzy functions, then $S \circ R$ is a perfect (strong) fuzzy function.*

Proof. a) Let R and S be strong fuzzy functions. For any $x, x' \in X$, $y, y' \in Y$, $z, z' \in Z$

$$T(R(x,y),S(y,z),R(x',y'),S(y',z'),E(x,x'))$$
$$\leq T(F(y,y'),S(y,z),S(y',z')) \leq G(z,z').$$

Therefore

$$T((S \circ R)(x,z),(S \circ R)(x',z'),E(x,x'))$$
$$= T(\sup_{y \in Y} T(R(x,y),S(y,z)), \sup_{y' \in Y} T(R(x,y'),S(y',z)),E(x,x'))$$
$$= \sup_{y,y' \in Y} T(R(x,y),S(y,z),(R(x,y'),S(y',z),E(x,x')) \leq G(z,z').$$

b) Let R and S be perfect fuzzy functions. For any $x, x' \in X$, $y, y' \in Y$, $z, z' \in Z$

(i) Extensionality.

$$T(R(x,y),S(y,z),E(x,x'),G(z,z'))$$
$$= T(R(x,y),S(y,z),E(x,x'),F(y,y),G(z,z'))$$
$$\leq T(R(x',y),S(y,z),G(z,z'))$$
$$= T(R(x',y),S(y,z),F(y,y),G(z,z'))$$
$$\leq T(R(x'y),S(y,z'))$$
$$\leq \sup_{y'' \in Y} T(R(x'y''),S(y'',z')) = (S \circ R)(x',z').$$

Therefore

$$T((S \circ R)(x,z),E(x,x'),G(z,z'))$$
$$= T(\sup_{y \in Y} T(R(x,y),S(y,z)),E(x,x'),G(z,z'))$$
$$= \sup_{y \in Y} T(R(x,y),S(y,z),E(x,x'),G(z,z')) \leq (S \circ R)(x',z').$$

(ii)$S \circ R$ is a partial fuzzy function.

$$T(R(x,y),S(y,z),R(x,y'),S(y'z'))$$
$$\leq T(F(y,y'),S(y,z),S(y',z')) \leq G(z,z').$$

The last inequality follows from the fact that since S is a perfect fuzzy function, it is also a strong fuzzy function. Therefore,

$$T((S \circ R)(x,z),(S \circ R)(x,z'))$$
$$= T(\sup_{y \in Y} T(R(x,y),S(y,z)), \sup_{y' \in Y} T(R(x,y'),S(y',z))$$
$$= \sup_{y,y' \in Y} T(R(x,y),S(y,z),R(x,y'),S(y',z)) \leq G(z,z').$$

(iii)$\forall x \in X\ \exists z \in Z$ such that $(S \circ R)(x,z) = 1$.
Given $x \in X$, there exists $y_1 \in Y$ and $z \in Z$ with $R(x,y_1) = 1$ and $R(y_1,z) = 1$.

$$\begin{aligned}(S \circ R)(x,z) &= \sup_{y \in Y} T(R(x,y), S(y,z)) \\ &\geq T(R(x,y_1), S(y_1,z)) = T(1,1) = 1.\end{aligned}$$

Definition 10.14. *Let E and F be T-indistinguishability operators on X and Y respectively, $x, x' \in X$ and $y, y' \in Y$.*

a) A perfect fuzzy function $R \in [0,1]^{X \times Y}$ is one to one if and only if

$$T(R(x,y), R(x',y)) \leq E(x,x').$$

b) A strong fuzzy function $R \in [0,1]^{X \times Y}$ is one to one if and only if

$$T(R(x,y), R(x',y'), F(y,y')) \leq E(x,x').$$

c) A perfect or strong fuzzy function is surjective if and only if

$$\forall y \in Y\ \exists x \in X \text{ such that } R(x,y) = 1.$$

d) A perfect or strong fuzzy function is bijective if and only if it is one to one and surjective.

Definition 10.15. *The inverse R^{-1} of a fuzzy relation R on $X \times Y$ is the fuzzy relation on $Y \times X$ defined for all $x \in X$, $y \in Y$ by*

$$R^{-1}(y,x) = R(x,y).$$

Proposition 10.16. *Let E and F be T-indistinguishability operators on X and Y respectively and R a fuzzy relation on $X \times Y$. R is a perfect (strong) fuzzy function if and only if R^{-1} is a perfect (strong) fuzzy function.*

Proposition 10.17. *If R is a bijective perfect fuzzy function, then*

$$R^{-1} \circ R = E \quad \text{and} \quad R \circ R^{-1} = F.$$

Proof. a) $R^{-1} \circ R = E$: For all $x, x' \in X$ and $y \in Y$,

$$T(R(x,y), R^{-1}(y,x')) = T(R(x,y), R(x',y)) \leq E(x,x')$$

and from this,

$$(R^{-1} \circ R)(x,x') = \sup_{y \in Y} T(R(x,y), R^{-1}(y,x')) \leq E(x,x').$$

On the other hand, there exists a function $f : X \to Y$ with $R(x,y) = F(f(x),y)$ and hence

$$\begin{aligned} E(x,x') &= T(E(x,x'), R(x', f(x'))) \\ &\leq R(x, f(x')) \\ &= T(R(x, f(x')), R(x', f(x'))) \\ &= T(R(x, f(x')), R^{-1}(f(x'), x')) \\ &\leq \sup_{y \in Y} T(R(x,y), R^{-1}(y, x')) \\ &= (R^{-1} \circ R)(x, x'). \end{aligned}$$

b) $R \circ R^{-1} = F$: Since R^{-1} is a perfect fuzzy function and $R = (R^{-1})^{-1}$, the proof of b) is the same as the proof of a).

10.3 Maximal Fuzzy Functions

Maximal fuzzy functions handle the greatest uncertainty in the sense that the granules are of the greatest possible size. Among them, perfect fuzzy functions are very natural, since in many cases we know, for example, that the image of an element x is y, which means that the fuzzy subset $R(x, \cdot)$ is normal and $R(x, y) = 1$.

Let us prove that the set of fuzzy functions contains maximal elements.

Proposition 10.18. *Let A be the set of partial fuzzy functions on $X \times Y$ with E and F T-indistinguishability operators on X and Y respectively. A has maximal elements.*

Proof. If $(R_i)_{i \in I}$ is a chain of partial fuzzy functions on $X \times Y$, then the fuzzy relation $R = \sup_{i \in I} R_i$ is a partial fuzzy function.

Applying Zorn's lemma, A has maximal elements.

Corollary 10.19. *Every partial fuzzy function on $X \times Y$ with E and F T-indistinguishability operators on X and Y respectively is contained in a maximal partial fuzzy function on $X \times Y$.*

Example 10.20. Let T be the Product t-norm, $X = \{a, b\}$, $Y = \{c, d\}$, E the T-indistinguishability operator on X defined by $E(a, b) = \frac{1}{2}$, F the T-indistinguishability operator on Y defined by $F(c, d) = \frac{1}{2}$ and R the partial fuzzy function on $X \times Y$ defined by $R(x, y) = \frac{1}{2}$ for all $x \in X$ $y \in Y$. R is not maximal and is contained in, for example, the maximal partial fuzzy function $\bar{R}$ with $\bar{R}(x, y) = \frac{1}{\sqrt{2}}$ $\forall x \in X$, $\forall y \in Y$.

It is trivial to see that $\bar{R}$ is a partial fuzzy function from X to Y. $\bar{R}$ is indeed maximal since if R' were a partial fuzzy function from X to Y with $R' > \bar{R}$, then there would exist an element of X and an element of Y (without loss of generality we can assume that they are a and c) with $R'(a, c) > \frac{1}{\sqrt{2}}$. But then $T(R'(a, c), R'(a, d))$ would be greater than $\frac{1}{2} = F(c, d)$ contradicting 10.2.1.

In fact, it is easy to prove that the fuzzy subsets $\bar{R}(a, \cdot)$ and $\bar{R}(b, \cdot)$ are maximal fuzzy points of Y. Next proposition generalizes this result.

Proposition 10.21. *Let R be a partial fuzzy function on $X \times Y$ with E and F T-indistinguishability operators on X and Y respectively. If for all $x \in X$ $R(x, \cdot)$ is a maximal fuzzy point of Y, then R is a maximal partial fuzzy function.*

Corollary 10.22. *Let R be a perfect fuzzy function on $X \times Y$ with E and F T-indistinguishability operators on X and Y respectively. Then for all $x \in X$ $R(x, y) = F(y_x, y)$.*

In other words, $R(x, \cdot)$ is a maximal fuzzy point of Y for every $x \in X$.

10.4 Classical Fuzzy Functions

Let us consider the special case of fuzzy function in which the equality on the domain is the crisp equality. The first attempts to fuzzify the idea of function between two sets assumed this condition in many cases [32]. The elements of the domain are considered crisp and completely distinguishable between them and there is only vagueness or uncertainty in their images.

Classical fuzzy functions appear in many contexts, since many times we know exactly the elements of our domain and are completely distinguishable. Then we only are uncertain about which exact image each elements of the domain has.

Definition 10.23. *A classical partial fuzzy function R is a partial fuzzy function on $X \times Y$ with E the classical equality on X ($E(x, x') = 0$ if $x \neq x'$) and F a T-indistinguishability operator on Y.*

Proposition 10.24. *A fuzzy subset R on $X \times Y$ is a classical partial fuzzy function on $X \times Y$ with F a T-indistinguishability operator on Y if and only if $R(x, \cdot)$ is a fuzzy point of Y with respect to F for all $x \in X$.*

Proof. Extensionality of R is

$$T(F(y, y'), R(x, y)) \leq R(x, y')$$

for all $x \in X$, $y, y' \in Y$, which is equivalent to say that $R(x, \cdot)$ is extensional with respect to F. The condition 10.2.1

$$T(R(x, y), R(x, y') \leq F(y, y')$$

also expresses that $R(x, \cdot)$ is a fuzzy point for all $x \in X$.

From a classical fuzzy function on $X \times Y$ we have a function $f : X \to \mathcal{F}(Y)$ where $\mathcal{F}(Y)$ is the set of fuzzy subsets of Y by simply defining $f(x) = R(x, \cdot)$.

Reciprocally,

Proposition 10.25. *Let $f : X \to Y^{[0,1]}$ be a function from X to the set $Y^{[0,1]}$ of fuzzy subsets of Y and R the fuzzy subset of $X \times Y$ defined by $R(x, y) = (f(x))(y)$ for all $x, y \in X \times Y$. R is a classical partial fuzzy function on $X \times Y$ with F a T-indistinguishability operator on Y if and only if for all $x \in X$ $R(x, \cdot)$ is a fuzzy point of Y.*

Corollary 10.26. *With the previous notations, R is a maximal classical partial fuzzy function if and only if the fuzzy points $R(x, \cdot)$ are maximal.*

This corollary allows us to construct maximal classical partial fuzzy functions containing a given classical partial fuzzy function.

Corollary 10.27. *Let R be a classical partial fuzzy function on $X \times Y$ with F a T-indistinguishability operator on Y. For all $x \in X$ let $\bar{R}(x, \cdot)$ be a maximal fuzzy point of Y containing $R(x, \cdot)$. Then $\bar{R}$ is a maximal partial classical fuzzy function on $X \times Y$ containing R.*

Example 10.28. Let $X = Y = \mathbb{R}$ be the real line, T the Łukasiewicz t-norm, F the T-indistinguishability operator on $\mathbb{R}$ defined by $F(x, y) = \max(0, 1 - |x - y|)$ for all $x, y \in \mathbb{R}$, $f : \mathbb{R} \to \mathbb{R}$ a function and R the classical partial fuzzy function on $[0, 1] \times [0, 1]$ defined by $R(x, y) = \max(0, \alpha - \alpha\,|y - f(x)|)$ for a given $\alpha \in [0, 1)$. R is not maximal and is contained in for example the classical partial maximal function $\bar{R}(x, y) = \max(0, 1 - |y - f(x)|)$, which is also perfect. The corresponding extensional function $f_{\bar{R}}$ is f.

Note that if the T-indistinguishability E on X is the classical equality, then every function $f : X \to Y$ is extensional. Also $\bar{R}(x, y)$ is maximal since it is perfect.

11

Indistinguishability Operators and Approximate Reasoning

In approximate reasoning, imprecise conclusions are inferred from imprecise premises. The typical way to do this is by using IF-THEN rules of the form

If x is A, then y is B.

A and B are modelled by fuzzy subsets μ_A and ν_B, respectively, and the conditions x is A and y is B are measured by $\mu_A(x)$ and $\nu_B(y)$.

This chapter presents two approaches to approximate reasoning based on indistinguishability operators.

The first one is based on extensionality and deals with the following scheme (Generalized Modus Ponens):

$$\begin{array}{c} \text{If } x \text{ is } A, \text{ then } y \text{ is } B \\ x \text{ is } A' \\ \hline y \text{ is } B' \end{array}$$

Although there are many ways to interpret the scheme, in the next section approximate reasoning will be based on the concept of proximity or similarity [16]:

If A' is close (or similar) to A, then B' must be close (or similar) to B.

The second approach is based on [80], where fuzzy functions are used to interpret approximate reasoning and Mamdani fuzzy controllers in their context.

Consider a family of fuzzy rules of the following form:

If x is A_i, then y is B_i $i = 1, 2, ..., n$

where the linguistic terms A_i and B_i are modelled by the fuzzy subsets μ_{A_i} of X and ν_{B_i} of Y.

These fuzzy rules are considered as patches of an unknown fuzzy function. From this fuzzy function, a control function is desired. Section 11.2 is devoted to this approach.

J. Recasens: Indistinguishability Operator, STUDFUZZ 260, pp. 189–199.
springerlink.com

11.1 Extensional Approach to Approximate Reasoning

Given a fuzzy IF-THEN rule

If x is A, then y is B,

we want to extend it to new rules of the form

If x is A', then y is B',

where A' is similar in some sense to A and A, A' are modelled by fuzzy subsets μ_A, μ of a universe of discourse X and B, B' by fuzzy subsets ν_B, ν of a universe of discourse Y.

The most usual way to infer B' from A' is using the compositional rule of inference (CRI).

Definition 11.1. *From the scheme*

$$\frac{\begin{array}{c}\text{If } x \text{ is } A, \text{ then } y \text{ is } B\\ x \text{ is } A'\end{array}}{y \text{ is } B'}$$

B' is obtained by the compositional rule of inference (CRI) in the following way

$$CRI(\mu)(y) = \nu(y) = \sup_{x \in X} T(R_{AB}(x, y), \mu(x))$$

for all $y \in Y$, where R_{AB} is a fuzzy relation (i.e. a fuzzy subset of $X \times Y$) and T a t-norm.

A number of fuzzy relations R_{AB} have been used to model the relation between A and B. The most common ones are

- $R_{AB}(x, y) = \min(\mu_A(x), \mu_B(y))$ (Mamdani)
- $R_{AB}(x, y) = T(\mu_A(x), \mu_B(y))$ (Mamdani with a t-norm T)
- $R_{AB}(x, y) = \overrightarrow{T}(\mu_A(x)|\mu_B(y))$ (Residuation based).

In an inference process based on the fuzzy IF-THEN rule 'If x is A, then y is B', we expect that if $A' \subseteq A''$, then the set consequence of A' be included in the consequence of A''.

Definition 11.2. *A map $\mathcal{C} : [0, 1]^X \to [0, 1]^Y$ is an inference operator if and only if it preserves the ordering:*

$$\mu \leq \mu' \Rightarrow \mathcal{C}(\mu) \leq \mathcal{C}(\mu').$$

Definition 11.3. *A map $\mathcal{C} : [0, 1]^X \to [0, 1]^Y$ is an extensional operator if and only if*

$$E_T(\mu, \mu') \leq E_T(\mathcal{C}(\mu), \mathcal{C}(\mu')).$$

(cf. Definition 10.4).

It happens that CRI is an extensional inference operator for any continuous t-norm and any relation R_{AB}.

Proposition 11.4. *For any continuous t-norm T and any fuzzy relation $R_{AB} : X \times Y \to [0,1]$ the CRI is an extensional inference operator.*

Proof. Monotonicity of T and sup assures that CRI is an inference operator.

In order to see that CRI is extensional we must prove that given $\mu, \mu' \in [0,1]^X$, we have $E_T(\mu, \mu') \leq E_T(\nu, \nu')$ where $\nu = \text{CRI}(\mu)$ and $\nu' = \text{CRI}(\mu')$.

Given $y \in Y$,

$$\begin{aligned}
\overrightarrow{T}(\nu(y)|\nu'(y)) &= \overrightarrow{T}(\text{CRI}(\mu)(y)|\text{CRI}(\mu')(y)) \\
&= \overrightarrow{T}(\sup_{x\in X} T(\mu(x), R_{AB}(x,y))| \sup_{x\in X} T(\mu'(x), R_{AB}(x,y))) \\
&= \inf_{x\in X} \overrightarrow{T}(T(\mu(x), R_{AB}(x,y))| \sup_{z\in X} T(\mu'(z), R_{AB}(z,y))) \\
&\geq \inf_{x\in X} \overrightarrow{T}(T(\mu(x), R_{AB}(x,y))|T(\mu'(x), R_{AB}(x,y))) \\
&\geq_{(*)} \inf_{x\in X} \overrightarrow{T}(\mu(x)|\mu'(x)) \\
&\geq \inf_{x\in X} \min\left(\overrightarrow{T}(\mu(x)|\mu'(x)), \overrightarrow{T}(\mu'(x)|\mu(x))\right) \\
&= \inf_{x\in X} E_(\mu(x), \mu'(x)) = E_T(\mu, \mu')
\end{aligned}$$

where (*) follows from Corollary 2.46.

Similarly we can prove

$$\overrightarrow{T}(\nu'(y)|\nu(y)) \geq E_T(\mu, \mu')$$

and therefore

$$\begin{aligned}
E_T(\nu(y), \nu'(y)) &= \min(\overrightarrow{T}(\nu(y)|\nu'(y)), \overrightarrow{T}(\nu'(y)|\nu(y))) \\
&\geq E_T(\mu, \mu').
\end{aligned}$$

Finally,

$$E_T(\nu, \nu') = \inf_{y\in Y} E_T(\nu(y), \nu'(y)) \geq E_T(\mu, \mu').$$

This result shows that CRI is an extensional inference operator despite the fuzzy relation R_{AB} used. In particular, it explains why the Mamdani fuzzy relation can be used in approximate reasoning and especially in fuzzy control.

Another example of an extensional inference operator is obtained from a crisp map f using the extension principle [143].

Definition 11.5. *Let $f : X \to Y$ be a crisp function between X and Y. f is extended to $f^* : [0,1]^X \to [0,1]^Y$ by the extension principle in the following way.*

$$f^*(\mu)(y) = \begin{cases} \sup_{x\in X, y=f(x)} \{\mu(x)\} & \text{if } \{f^{-1}(y)\} \neq \emptyset \\ 0 & \textit{otherwise.} \end{cases}$$

Proposition 11.6. *The map $f^* : [0,1]^X \to [0,1]^Y$ obtained from $f : X \to Y$ by the extension principle is an extensional inference operator for every continuous t-norm T.*

Proof. It is trivial to prove that f^* is an inference operator.
Let us prove that f^* is extensional.
Given $y \in Y$,

$$\begin{aligned}\overrightarrow{T}(f^*(\mu)(y)|f^*(\mu')(y)) &= \overrightarrow{T}(\sup_{x\in\{f^{-1}(y)\}} \mu(x)| \sup_{x\in\{f^{-1}(y)\}} \mu'(x)) \\ &= \inf_{x\in\{f^{-1}(y)\}} \overrightarrow{T}(\mu(x)| \sup_{z\in\{f^{-1}(y)\}} \mu'(z)) \\ &\geq \inf_{x\in\{f^{-1}(y)\}} \overrightarrow{T}(\mu(x)|\mu'(x)) \\ &\geq \inf_{x\in X} \overrightarrow{T}(\mu(x)|\mu'(x)) \\ &\geq \inf_{x\in X} E_T(\mu(x),\mu'(x)) = E_T(\mu,\mu').\end{aligned}$$

Similarly we can prove

$$\overrightarrow{T}(f^*(\mu')(y)|f^*(\mu)(y)) \geq E_T(\mu,\mu')$$

and therefore

$$E_T(f^*(\mu)(y), f^*(\mu')(y)) \geq E_T(\mu,\mu').$$

Finally,

$$E_T(f^*(\mu), f^*(\mu')) \geq E_T(\mu,\mu').$$

Let us introduce the natural inference operator (NIO) that is optimal in the sense of Theorem 11.8.

Definition 11.7. *Given the rule 'If x is A, then y is B', with μ_A and ν_B fuzzy subsets of X and Y respectively, the natural inference operator (NIO)*

$$\begin{aligned}\mathcal{C}_{AB} : [0,1]^X &\to [0,1]^Y \\ \mu &\to \mathcal{C}_{AB}(\mu) = \nu\end{aligned}$$

is defined by

$$\mathcal{C}_{AB}(\mu)(y) = \nu(y) = \overrightarrow{T}(\inf_{x\in X} \overrightarrow{T}(\mu(x)|\mu_A(x))|\nu_B(y)).$$

Theorem 11.8. *The NIO $\mathcal{C}_{AB}$ satisfies*

1. *$\mathcal{C}_{AB}$ is an inference operator.*
2. *$\nu_B \leq \mathcal{C}_{AB}(\mu)$ for all $\mu \in [0,1]^X$. Moreover, if $\mu \leq \mu_A$, then $\mathcal{C}_{AB}(\mu) = \nu_B$.*
3. *$\mathcal{C}_{AB}$ interpolates the rule If x is A, then y is B (i.e. $\mathcal{C}_{AB}(\mu_A) = \nu_B$).*
4. *$\mathcal{C}_{AB}$ is extensional.*

5. $\mathcal{C}_{AB}$ *is the least specific (the greatest) operator satisfying the preceding properties.*

Proof

1. It is a consequence of the monotonicity of the t-norm T and the infimum.
2. For any $x, y \in [0,1]$, $\overrightarrow{T}(x|y) \geq y$.
   Therefore, given $\mu \in [0,1]^X$, for all $y \in Y$

$$\mathcal{C}_{AB}(\mu)(y) = \nu(y) = \overrightarrow{T}(\inf_{x \in X} \overrightarrow{T}(\mu(x)|\mu_A(x))|\nu_B(y)) \geq \nu_B(y).$$

   If $\mu \leq \mu_A$,

$$\begin{aligned} \mathcal{C}_{AB}(\mu)(y) &= \nu(y) \\ &= \overrightarrow{T}(\inf_{x \in X} \overrightarrow{T}(\mu(x)|\mu_A(x))|\nu_B(y)) \\ &= \overrightarrow{T}(1|\nu_B(y) = \nu_B(y) \end{aligned}$$

   for all $y \in T$.
3. It follows from 2.
4. For $\mu, \mu' \in [0,1]^X$, let $\nu = \mathcal{C}_{AB}(\mu)$ and $\nu' = \mathcal{C}_{AB}(\mu')$.
   We must prove

$$E_T(\mu, \mu') \leq E_T(\nu, \nu').$$

   Given $y \in Y$,

$$\begin{aligned} &\overrightarrow{T}(\nu'(y)|\nu(y)) \\ &= \overrightarrow{T}(\overrightarrow{T}(\inf_{x \in X} \overrightarrow{T}(\mu(x)|\mu_A(x))|\nu_B(y))|\overrightarrow{T}(\inf_{x \in X} \overrightarrow{T}(\mu'(x)|\mu_A(x))|\nu_B(y))) \\ &\geq_{(*)} \overrightarrow{T}(\inf_{x \in X} \overrightarrow{T}(\mu'(x)|\mu_A(x))|\inf_{x \in X} \overrightarrow{T}(\mu(x)|\mu_A(x))) \\ &= \inf_{x \in X} \overrightarrow{T}(\inf_{z \in X} \overrightarrow{T}(\mu'(z)|\mu_A(z))|\overrightarrow{T}(\mu(x)|\mu_A(x))) \\ &\geq \inf_{x \in X} \overrightarrow{T}(\overrightarrow{T}(\mu'(x)|\mu_A(x))|\overrightarrow{T}(\mu(x)|\mu_A(x))) \\ &\geq_{(**)} \inf_{x \in X} \overrightarrow{T}(\mu(x)|\mu'(x)) \\ &\geq \inf_{x \in X} \min(\overrightarrow{T}(\mu(x)|\mu'(x)), \overrightarrow{T}(\mu'(x)|\mu(x))) \\ &= \inf_{x \in X} E_T(\mu(x), \mu'(x)) = E_T(\mu, \mu') \end{aligned}$$

   where (*) and (**) are consequence of Corollary 2.46.
   Similarly we can prove

$$\overrightarrow{T}(\nu(y)|\nu'(y)) \geq E_T(\mu, \mu').$$

Therefore

$$E_T(\nu(y), \nu'(y)) = \min(\overrightarrow{T}(\nu(y)|\nu'(y)), \overrightarrow{T}(\nu'(y)|\nu(y))) \geq E_T(\mu, \mu').$$

Finally,

$$E_T(\nu, \nu') = \inf_{y \in Y} E_T(\nu(y), \nu'(y)) \geq E_T(\mu, \mu').$$

5. Let $\mathcal{D} : [0,1]^X \to [0,1]^Y$ be an operator satisfying the preceding properties. Given $\mu \in [0,1]^X$ we want to prove that $\mathcal{D}(\mu) \leq \mathcal{C}_{AB}(\mu)$.

   Let μ' be the fuzzy subset of X defined for all $x \in X$ by $\mu'(x) = \max(\mu_A(x), \mu(x))$.

   Since $\mu_A \leq \mu'$,

$$\begin{aligned} E_T(\mu', \mu_A) &= \inf_{x \in X} E_T(\mu'(x), \mu_A(x)) \\ &= \inf_{x \in X} \overrightarrow{T}(\mu'(x)|\mu_A(x)). \end{aligned}$$

Since $\mathcal{D}$ satisfies 2., 3. and 4.,

$$\begin{aligned} &\inf_{y \in Y} \overrightarrow{T}(\mathcal{D}(\mu')(y)|\nu_B(y)) \\ &= \inf_{y \in Y} \min(\overrightarrow{T}(\mathcal{D}(\mu')(y)|\nu_B(y)), \overrightarrow{T}(\nu_B(y)|\mathcal{D}(\mu')(y))) \\ &= \inf_{y \in Y} E_T(\mathcal{D}(\mu')(y), \nu_B(y)) \\ &= E_T(\mathcal{D}(\mu')|\nu_B) \\ &= E_T(\mathcal{D}(\mu')|\mathcal{D}(\mu_A)) \\ &\geq E_T(\mu', \mu_A) \\ &= \inf_{x \in X} \overrightarrow{T}(\mu'(x)|\mu_A(x)). \end{aligned}$$

Therefore, for all $y \in Y$,

$$\overrightarrow{T}(\mathcal{D}(\mu')(y)|\nu_B(y)) \geq \inf_{x \in X} \overrightarrow{T}(\mu'(x)|\mu_A(x)).$$

As a consequence of Lemma 2.41,

$$\begin{aligned} \mathcal{D}(\mu')(y) &\leq \sup\{\alpha \in [0,1] \text{ such that } \overrightarrow{T}(\alpha|\nu_B(y)) \geq \inf_{x \in X} \overrightarrow{T}(\mu'(x)|\mu_A(x)\} \\ &= \overrightarrow{T}(\inf_{x \in X} \overrightarrow{T}(\mu'(x)|\mu_A(x))|\nu_B(y)) \end{aligned}$$

and hence

$$\mathcal{D}(\mu') \leq \mathcal{C}_{AB}(\mu').$$

But since $\mu' = \max(\mu, \mu_A)$ and $\mathcal{C}_{AB}$ is an inference operator,

$$\begin{aligned} \mathcal{C}_{AB}(\mu') &= (\mu \vee \mu_A) \\ &= \mathcal{C}_{AB}(\mu) \vee \mathcal{C}_{AB}(\mu_A) = \mathcal{C}_{AB}(\mu). \end{aligned}$$

Also since $\mathcal{D}$ satisfies 1. and $\mu \leq \mu'$, we have $\mathcal{D}(\mu) \leq \mathcal{D}(\mu')$ obtaining

$$\mathcal{D}(\mu) \leq \mathcal{D}(\mu') \leq \mathcal{C}_{AB}(\mu') = \mathcal{C}_{AB}(\mu).$$

11.2 Approximate Reasoning, Fuzzy Control and Fuzzy Functions

In this section, the preceding ideas are specialized to fuzzy control.

Let us first recall how fuzzy control works.

(1)There is a finite family of fuzzy rules

$\mathcal{R}_i$: If x is A_i, then y is $B_i \quad i = 1, 2, ..., n$

where A_i and B_i are modelled by fuzzy subsets μ_{A_i} and ν_{B_i} of the universes X and Y respectively (usually subsets of $\mathbb{R}$ or $\mathbb{R}^n$), x is A_i and y is B_i are expressed by $\mu_{A_i}(x)$ and $\nu_{B_i}(y)$ respectively and for every rule $\mathcal{R}_i$ the fuzzy relation R_i that relates x and y is generated by μ_{A_i} and ν_{B_i}. The most usual obtained relations are

a) $R_{T_i}(x, y) = T(\mu_{A_i}(x), \nu_{B_i}(y))$ (Mamdani type)
or
b) $R_{\rightarrow_i}(x, y) = \overrightarrow{T}(\mu_{A_i}(x)|\nu_{B_i}(y))$ (generated by the residuation of a t-norm).

(2)Given $x \in X$, for every rule $\mathcal{R}_i$ a fuzzy subset $\nu_{x,i}$ of Y is computed by

$$\nu_{x,i}(y) = R_i(x, y). \tag{1}$$

If R_{T_i} are of Mamdani type,this means that

$$\nu_{x,i}(y) = T(\mu_{A_i}(x), \nu_{B_i}(y))$$

and for the residuation type relations $R_{\rightarrow_i}$,

$$\nu_{x,i}(y) = \overrightarrow{T}(\mu_{A_i}(x)|\nu_{B_i}(y)).$$

(3)The fuzzy rules are combined to obtain a final fuzzy subset ν_x in the following way

$$\nu_x(y) = \max\{\nu_{x,i}, \ i = 1, 2, ..., n\} \tag{2}$$

for the Mamdani type relations and

$$\nu_x(y) = \min\{\nu_{x,i}, \ i = 1, 2, ..., n\} \tag{3}$$

for the residuation type relations.

They correspond to considering the fuzzy relations

$$R_T = \max\{R_{T_i},\ i = 1, 2, ..., n\}$$

and

$$R_{\rightarrow} = \min\{R_{\rightarrow_i},\ i = 1, 2, ..., n\}$$

respectively.

(4)Finally, a defuzzification method finds a precise $y_x \in Y$ and a control map $f : x \rightarrow f(x) = y_x$ is obtained.

In this context, the fuzzy rules are considered as patches of a fuzzy function and from these patches a crisp function is wanted.

Suppose now that T-indistinguishability operators E and F are defined on X and Y respectively. In this case the fuzzy subsets μ_{A_i} and ν_{B_i} should be extensional with respect to E and F respectively since only extensional fuzzy subsets can be observed when the indistinguishability operator is taken into account (see Chapter 3). In this case, the relations R_T and $R_{\rightarrow}$ are extensional.

Proposition 11.9. *Let E and F be T-indistinguishability operators on X and Y respectively and the fuzzy subsets μ_{A_i}, ν_{B_i}, $i = 1, 2, ..., n$ extensional with respect to E and F respectively. Then the fuzzy relations R_T and $R_{\rightarrow}$ are extensional with respect to E and F.*

Proof. R_T is trivially extensional since the fuzzy subsets are.

Let us prove that $R_{\rightarrow}$ is also extensional.

$$\begin{aligned}\nu_{B_i}(y) &\geq T(\mu_{A_i}(x), \overrightarrow{T}(\mu_{A_i}(x)|\nu_{B_i}(y))) \\ &\geq T(\mu_{A_i}(x'), E(x, x'), \overrightarrow{T}(\mu_{A_i}(x)|\nu_{B_i}(y))).\end{aligned}$$

which is equivalent to

$$T(E(x, x'), \overrightarrow{T}(\mu_{A_i}(x)|\nu_{B_i}(y))) \leq \overrightarrow{T}(\mu_{A_i}(x')|\nu_{B_i}(y)).$$

Taking the minimum over all $i = 1, 2, ..., n$ we get

$$T(E(x, x'), R_{\rightarrow}(x, y)) \leq R_{\rightarrow}(x', y). \quad (4)$$

$$\begin{aligned}T(R_{\rightarrow}(x, y), F(y, y')) &\leq \min_{i=1,2,\ldots,n} (T(\overrightarrow{T}(\mu_{A_i}(x)|\nu_{B_i}(y)), F(y, y'))) \ (5) \\ &\leq \min_{i=1,2,\ldots,n} \overrightarrow{T}(\mu_{a_i}(x)|T(\nu_{B_i}(y), F(y, y'))) \\ &\leq \min_{i=1,2,\ldots,n} \overrightarrow{T}(\mu_{a_i}(x)|\nu_{B_i}(y')).\end{aligned}$$

(4) and (5) prove the extensionality of $R_{\rightarrow}$.

Also every element x must be replaced by the less specific extensional fuzzy subset containing it in order to be observable with respect to E. This subset is $\phi_E(\{x\}) = \mu_x = E(x, \cdot)$ the column associated to x.

If the fuzzy subsets μ_{A_i} satisfy the conditions of Proposition 3.69, then they are also columns of E.

Proposition 11.10. *Let E and F be T-indistinguishability operators on X and Y respectively, the fuzzy subset μ_A a column of E and $x' \in X$. The fuzzy subset ν of Y obtained by the column $\mu_{x'}$ associated to x' by CRI when $R_{AB} = T(\mu_A, \nu_B)$ or $R_{AB} = \overrightarrow{T}(\mu_A|\nu_B)$ coincide with (1) ($\nu(y) = R_{AB}(x', y)$)).*

Proof

- Mamdani type:

$$\begin{aligned} CRI(\mu_{x'})(y) &= \nu(y) = \sup_{z\in X} T(\mu_{x'}(z), R_{AB}(z, y)) \\ &= \sup_{z\in X} T(E(x', z), E(x, z), \nu(y)) \\ &=_{(*)} T(E(x, x'), \nu(y)) \\ &= T(\mu_x(x'), \nu(y)) = R_{AB}(x', y). \end{aligned}$$

  (*) follows from the T-transitivity of E:

$$T(E((x', z), E(x, z)) \leq E(x, x')$$

  and for $z = x'$ equality holds.
- Residuation type:

$$\begin{aligned} CRI(\mu_{x'})(y) = \nu(y) &= \sup_{z\in X} T(\mu_{x'}(z), R_{AB}(z, y)) \\ &= \sup_{z\in X} T(E(x', z), \overrightarrow{T}(E(x, y)|\nu(y)). \end{aligned}$$

  Taking $z = x'$,

$$\begin{aligned} &\sup_{z\in X} T(E(x', z), \overrightarrow{T}(E(x, y)|\nu(y)) \\ &\quad \geq \overrightarrow{T}(E(x, x')|\nu(y) = R_{AB}(x', y). \end{aligned}$$

  On the other hand,

$$\begin{aligned} &\sup_{z\in X} T(E(x', z), \overrightarrow{T}(E(x, y)|\nu(y)) \\ &\quad \leq_{(*)} \sup_{z\in X} \overrightarrow{T}(\overrightarrow{T}(E(x', z)|E(x, z)|\nu(y)) \\ &\quad \leq_{(**)} \sup_{z\in X} \overrightarrow{T}(E(x, x')|\nu(y)) \\ &\quad \leq \overrightarrow{T}\mu_x(x')|\nu(y)) = R_{AB}(x', y). \end{aligned}$$

(*) follows from Lemma 2.42.
(**) follows from the T-transitivity of E and from the fact that $\overrightarrow{T}$ is non increasing with respect to the first variable.

Corollary 11.11. *Let E and F be T-indistinguishability operators on X and Y respectively, the fuzzy subsets μ_{A_i} columns of E and $x' \in X$. The fuzzy subset ν of Y obtained by the column $\mu_{x'}$ associated to x' by CRI with R_T and with $R_{\rightarrow}$ coincide with (2) and (3) respectively.*

These results justify the outputs ν obtained in fuzzy control after the process of fuzzification (obtaining $\phi_E(\{x'\}) = E(x', \cdot)$) and applying the CRI afterward when there is a T-indistinguishability operator E on X and the fuzzy subsets μ_{A_i} are columns of E (i.e. they are the fuzzification of precise values on X).

Proposition 11.12. *Let E, F be two T-indistinguishability operators on X and Y respectively such that the fuzzy subsets μ_{A_i} and ν_{B_i} are the columns associated to the points x_i and y_i respectively. If the partial function $f(x_i) = y_i$ is extensional with respect to E and F, then R_T is a partial fuzzy function and $R_T = R_f$ where $R_f(x, y) = \sup_{i=1,2,...,n} T(E(x, x_i), F(y, y_i))$ (Proposition 10.8.).*

Proof.

$$\begin{aligned} R_T(x, y) &= \sup_{i=1,2,\ldots,n} T(\mu_{A_i}, \nu_{B_i}) \\ &= \sup_{i=1,2,\ldots,n} T(E(x, x_i), F(y, y_i)) \\ &= R_f(x, y). \end{aligned}$$

Theorem 11.13. *Let E, F be two T-indistinguishability operators on X and Y respectively such that the fuzzy subsets μ_{A_i} and ν_{B_i} are the columns associated to the points x_i and y_i respectively and $f : X \rightarrow Y$ an extensional function with respect to E and F. Let f_D be the partial function with domain $D = \{x_1, x_2, ..., x_n\}$ defined as the restriction of f to D (i.e. $f_D(x_i) = f(x_i)$). Then*

$$R_T = R_{f_D} \leq R_f \leq R_{\rightarrow}.$$

Proof. The first equality has been proved in Proposition 11.12. The first inequality is obvious. Let us prove the last inequality.

$$T(E(x, x_i), F(y, f(x)) \leq T(F(f(x), f(x_i)), F(y, f(x))) \leq F(y, f(x_i))$$

or

$$F(y, f(x)) \leq \overrightarrow{T}(E(x, x_i)|F(y, f(x_i)) = \overrightarrow{T}(\mu_{A_i}(x)|\nu_{B_i}(y)).$$

Therefore

$$R_f(x, y) = F(y, f(x)) \leq \min_{i=1,2,\ldots,n} \overrightarrow{T}(\mu_{A_i}(x)|\nu_{B_i}(y)) = R_{\rightarrow}.$$

The last theorem gives upper and lower bounds to the fuzzy control function f. Indeed, from the rules a partial function f_D is obtained and the control function f must interpolate the values of f_D. Then R_T and $R_{\rightarrow}$ are bounds to the partial fuzzy function generated by the desired control function f.

12

Vague Groups

In the crisp case, if $(G, \circ)$ is a set with an operation $\circ : G \times G \rightarrow G$ and $\sim$ is an equivalence relation on G, then $\circ$ is compatible with $\sim$ if and only if

$$a \sim a' \text{ and } b \sim b' \text{ implies } a \circ b \sim a' \circ b'.$$

In this case, an operation $\tilde{\circ}$ can be defined on $\overline{G} = G/\sim$ by

$$\overline{a}\tilde{\circ}\overline{b} = \overline{a \circ b}$$

where $\overline{a}$ and $\overline{b}$ are the equivalence classes of a and b with respect to $\sim$.

Demirci generalized this idea to the fuzzy framework by introducing the concept of vague algebra, which basically consists of fuzzy operations compatible with given indistinguishability operators [37].

As an example of vague algebra, this chapter will examine the idea of vague groups and how they relate to fuzzy subgroups.

Fuzzy subgroups were introduced by Rosenfeld [120] as a natural generalization of the subgroup concept and have since been widely studied. Demirci later on introduced T-vague groups by forcing the operation of the group to be compatible with a given indistinguishability operator E.

In this chapter, the notion of T-vague group will be introduced, and it will be shown that T-vague groups can be viewed as the natural fuzzy generalization of quotient groups. When E separates points, then a T-vague group can be thought of as a quotient group of a given group $(G, \circ)$ by a normal fuzzy subgroup μ of G, with its core consisting of only the identity element of G.

There is a close relationship between fuzzy subgroups and T-vague groups. Indeed, we can firstly associate a T-indistinguishability operator $E_{(\mu)}$ with every fuzzy subgroup μ. Normal fuzzy subgroups of $(G, \circ)$ are defined as fuzzy subgroups μ satisfying $\mu(a \circ b) = \mu(b \circ a) \forall a, b \in X$ (Definition 12.22) and the most interesting result relating μ and $E_{\mu)}$ is Theorem 12.27, which, roughly speaking, states that the operation of the group is compatible with the

J. Recasens: Indistinguishability Operator, STUDFUZZ 260, pp. 201–215.
springerlink.com

T-indistinguishability operator $E_{\mu)}$ if and only if μ is a normal fuzzy subgroup of G.

Fuzzy subgroups and T-vague groups are related by the fact that given a normal fuzzy subgroup μ of G, G with the operation $\tilde{\circ}(a,b,c) = E(a \circ b, c)$ is a T-vague group and, reciprocally, for every T-vague group there exists a T-indistinguishability operator E that is invariant under translations such that $\tilde{\circ}(a,b,c) = E(a \circ b, c)$. It will be proved that a T-indistinguishability operator E is the operator associated with a normal fuzzy subgroup if and only if E is invariant under translations, meaning that all its classes differ only by a translation factor. Therefore, we will have natural bijections between the sets of normal fuzzy subgroups, T-indistinguishability operators invariant under translations and T-vague groups of a group $(G, \circ)$. In particular, T-vague groups can be thought of as the fuzzy counterparts of crisp quotient groups.

Section 12.4 is devoted to the case of the real line $(\mathbb{R}, +)$. Given a continuous Archimedean t-norm or the minimum t-norm, the normal fuzzy subgroups that also are fuzzy numbers are completely characterized, as well as the admissible T-indistinguishability operators and their corresponding T-vague groups. Two examples on the chapter show that these kind of results can give insight to some useful tools in fuzzy systems:

- In Example 12.46 triangular fuzzy numbers are seen as elements of $\mathbb{R}/\mu$ the quotient of the real line modulo a triangular symmetric fuzzy number centered at 0, which is in fact a normal fuzzy subgroup of $\mathbb{R}$ with respect to the Łukasiewicz t-norm.
- Example 12.47 develops the idea of being a vague multiple of a given integer number.

12.1 Fuzzy Subgroups and T-Vague Groups

This section contains some definitions and properties related to fuzzy subgroups and T-vague groups. Some properties of T-vague operations are also stated so that T-vague groups could be set in their context.

Definition 12.1. *Let $(G, \circ)$ be a group and μ a fuzzy subset of G. μ is a T-fuzzy subgroup of G if and only if*

$$T(\mu(a), \mu(b)) \leq \mu(a \circ b^{-1}) \ \forall a, b \in X.$$

Proposition 12.2. *Let $(G, \circ)$ be a group, e its identity element and μ a fuzzy subset of G such that $\mu(e) = 1$. Then μ is a T-fuzzy subgroup of G if and only if $\forall a, b \in X$ the following properties hold*

a) $\mu(a) = \mu(a^{-1})$
b) $T(\mu(a), \mu(b)) \leq \mu(a \circ b)$.

Proof

$\Rightarrow$)

a) If μ is a T-fuzzy subgroup with $\mu(e) = 1$, then

$$\mu(a) = T(\mu(e), \mu(a)) \leq \mu(a^{-1}).$$

By symmetry, $\mu(a) = \mu(a^{-1})$ holds.

b) $T(\mu(a), \mu(b)) = T(\mu(a), \mu(b^{-1})) \leq \mu(a \circ b)$.

$\Leftarrow$) $T(\mu(a), \mu(b)) = T(\mu(a), \mu(b^{-1})) \leq \mu(a \circ b^{-1})$.

Definition 12.3. *The core H of a fuzzy subset μ of a set G is the set of elements a of G such that $\mu(a) = 1$.*

Proposition 12.4. *Let $(G, \circ)$ be a group, e its identity element and μ a fuzzy subgroup of X such that $\mu(e) = 1$. Then the core H of μ is a subgroup of G.*

Proof. Let $a, b \in H$.

$$1 = T(\mu(a), \mu(b)) = T(\mu(a), \mu(b^{-1})) \leq \mu(a \circ b^{-1})$$

and therefore $a \circ b^{-1} \in H$.

T-vague algebras were introduced by Demirci in [35] considering fuzzy operations compatible with given T-fuzzy equalities and an extensive study of vague operations and T-vague groups can be found in [34],[35].

Definition 12.5. *A fuzzy binary operation on a set G is a map $\tilde{\circ} : G \times G \times G \to [0, 1]$.*

$\tilde{\circ}(a, b, c)$ is interpreted as the degree in which c is $a \circ b$.

Definition 12.6. *Let E be T-indistinguishability operators on G. A vague binary operation on G is a perfect fuzzy map $\tilde{\circ}$ from $G \times G$ to G where in $G \times G$ the T-indistinguishability operator $G(E, E)$ defined for all $a, a', b, b' \in X$ by*

$$G(E, E)((a, b), (a', b')) = T(E(a, a'), E(b, b'))$$

is considered.

In a more explicit way, Definition 12.6 states that

a) $T(\tilde{\circ}(a, b, c), E(a, a'), E(b, b'), E(c, c')) \leq \tilde{\circ}(a', b', c')$.

b) $T(\tilde{\circ}(a, b, c), \tilde{\circ}(a, b, c')) \leq E(c, c')$.

c) For all $a, b \in G$ there exists $c \in G$ such that $\tilde{\circ}(a, b, c) = 1$.

These conditions fuzzify the idea of compatibility between a binary operation and an equivalence relation on a set.

- Condition a) states that if c is the (vague) result of operating a and b and a', b', c' are indistinguishable from a, b, c, then c' is the vague result of operating a' and b'.
- b) asserts that if c and c' are vague results of the operation $a \circ b$, then they are indistinguishable.
- c) says that for every a and b there exists c that is exactly the result of operating a with b.

Definition 12.7. *Let $\tilde{\circ}$ be a T-vague binary operation on G with respect to a T-indistinguishability operator E on G. Then $(G, \tilde{\circ})$ is a T-vague group if and only if it satisfies the following properties.*

1. *Associativity.* $\forall a, b, c, d, m, q, w, \in X$
$$T(\tilde{\circ}(b, c, d), \tilde{\circ}(a, d, m), \tilde{\circ}(a, b, q), \tilde{\circ}(q, c, w)) \leq E(m, w)).$$
2. *Identity. There exists a (two sided) identity element $e \in G$ such that*
$$T(\tilde{\circ}(e, a, a), \tilde{\circ}(a, e, a)) = 1$$
for each $a \in G$.
3. *Inverse. For each $a \in G$ there exists a (two-sided) inverse element $a^{-1} \in G$ such that*
$$T(\tilde{\circ}(a^{-1}, a, e), \tilde{\circ}(a, a^{-1}, e)) = 1.$$

A T-vague group is Abelian or commutative if and only if

$$\forall a, b, m, w \in G, T((\tilde{\circ}(a, b, m), \tilde{\circ}(b, a, w))) \leq E(m, w)).$$

For a given T-vague group $(G, \tilde{\circ})$, the identity and the inverse element a^{-1} of each $a \in G$ are not unique in general.

Proposition 12.8. *Let $(G, \tilde{\circ})$ be a T-vague group with respect to E. Then,*

1. *If e, e' are two identities of $(X, \tilde{\circ})$, then $E(e, e') = 1$.*
2. *If b, b' are two inverse elements of $a \in G$ with respect to an identity element of e of $(G, \tilde{\circ})$, then $E(b, b') = 1$.*
3. *If $E(a_1, a_2) = 1$, then $b \in G$ is an inverse element of $a_1 \in G$ with respect to an identity element e of $(G, \tilde{\circ})$ if and only if b is an inverse element of $a_2 \in G$ with respect to e.*
4. *If b is an inverse element of $a \in G$ with respect to a given identity element e of $(G, \tilde{\circ})$, then b is an inverse element of a with respect to all identity elements of $(G, \tilde{\circ})$.*
5. *If b, b' are inverse elements of $a \in X$ with respect to the identity elements e, e' respectively, then $E(b, b') = 1$.*
6. *If E separates points, then the identity and the inverse elements of $(G, \tilde{\circ})$ are unique.*

Proof

1. Since e and e' are identity elements of the T-vague group, $\tilde{\circ}(e, e', e) = 1$ and $\tilde{\circ}(e, e', e') = 1$.
   $1 = T(\tilde{\circ}(e, e', e), \tilde{\circ}(e, e', e'), E(e, e), E(e', e')) \leq E(e, e')$ and therefore $E(e, e') = 1$.
2. Due to the associativity of $\tilde{\circ}$,

$$1 = T(\tilde{\circ}(b, a, e), \tilde{\circ}(e, b', b'), \tilde{\circ}(a, b', e), \tilde{\circ}(b, e, b)) \leq E(b', b).$$

3.

$$1 = T(\tilde{\circ}(a_1, b, e), E(a_1, a_2)) \leq \tilde{\circ}(a_2, b, e).$$

   Also,

$$1 = T(\tilde{\circ}(b, a_1, e), E(a_1, a_2)) \leq \tilde{\circ}(b, a_2, e).$$

4.

$$1 = \tilde{\circ}(b, a, e) = T(\tilde{\circ}(b, a, e), E(e, e')) \leq \tilde{\circ}(b, a, e').$$

   Also,

$$1 = \tilde{\circ}(a, b, e) = T(\tilde{\circ}(a, b, e), E(e, e')) \leq \tilde{\circ}(a, b, e').$$

5. It is a straightforward consequence of 1. and 4.
6. 1. and 2. assure the uniqueness of the identity and the inverse elements respectively.

Let us recall that if E is a T-indistinguishability operator on a set G, then the (crisp) relation $\sim_E$ on G defined by $a \sim_E b$ if and only if $E(a, b) = 1$ is an equivalence relation. The fuzzy relation $\bar{E}$ on $G/\sim_E$ defined by $\bar{E}(\bar{a}, \bar{b}) = E(a, b)$ is a T-indistinguishability operator that separates points.

Proposition 12.9. *Let $(G, \tilde{\circ})$ be a T-vague group with respect to a T-indistinguishability operator E. In the quotient set $G/\sim_E$ let us consider the operation $\bar{\circ}$ defined for all $\bar{a}, \bar{b}, \bar{c} \in G/\sim_E$ by $\bar{a}\bar{\circ}\bar{b} = \bar{c}$ if and only if $\tilde{\circ}(a, b, c) = 1$. Then $\bar{\circ}$ is well defined.*

Proof. We need to show that
a) If $a \sim_E a', b \sim_E b', c \sim_E c'$ and $\tilde{\circ}(a, b, c) = 1$, then $\tilde{\circ}(a', b', c') = 1$.
b) If $a \sim_E a', b \sim_E b', \tilde{\circ}(a, b, c) = 1$ and $\tilde{\circ}(a', b', c') = 1$, then $c \sim_E c'$.
To prove a) is trivial using the transitivity of $\tilde{\circ}$ and b) is trivial since $\tilde{\circ}$ is a T-vague operation.

Corollary 12.10. *$(G/\sim_E, \bar{\circ})$ is a group.*

Corollary 12.11. *With the same notations as before $(G/\sim_E, \bar{\tilde{\circ}})$ is a T-vague group with respect to $\bar{E}$ where $\bar{\tilde{\circ}}(\bar{a}, \bar{b}, \bar{c}) = \tilde{\circ}(a, b, c)$ $\forall a, b, c \in G$.*

Therefore from every T-vague group we can define another T-vague group with respect to a separable T-indistinguishability operator and go back to the definition of T-vague group given in [34], [35] where separability of the T-indistinguishability operator was required.

12.2 Normal Fuzzy Subgroups and T-Indistinguishability Operators

In the crisp case, given a subgroup H of a group $(G, \circ)$ the relation $\sim$ on G defined by $a \sim b$ if and only if $a \circ b^{-1} \in H$ is an equivalence relation. The operation $\circ$ of G is compatible with $\sim$ if and only if H is a normal subgroup of G.

In this Section we will generalize these results to T-fuzzy subgroups and T-indistinguishability operators.

Definition 12.12. *Let $\circ$ be a binary operation on G, and E a T-indistinguishability operator on G. E is regular with respect to $\circ$ if and only if*

$$E(a,b) \leq \bigwedge\nolimits_{c\in G} E(a \circ c, b \circ c) \wedge E(c \circ a, c \circ b), \ \forall a, b \in G.$$

Definition 12.13. *Let $\circ$ be a binary operation on G and E a T-indistinguishability operator on G. E is invariant under translations with respect to $\circ$ if and only if*

a)

$$E(a,b) = E(c \circ a, c \circ b) \text{ (left invariant)}$$

and

b)

$$E(a,b) = E(a \circ c, b \circ c) \text{ (right invariant)},$$

$\forall a, b, c \in G$.

Proposition 12.14. *Let $(G, \circ)$ be a group and E a T-indistinguishability operator on G. Then E is regular with respect to $\circ$ if and only if E is invariant under translations with respect to $\circ$.*

Proof. $\Rightarrow$) Suppose that E is regular with respect to $\circ$. Then, we have

$$E(a,b) \leq \bigwedge\nolimits_{c\in X} E(a \circ c, b \circ c) \wedge E(c \circ z, c \circ b),$$

i.e.,

$$E(a,b) \leq E(a \circ c, b \circ c) \text{ and } E(a,b) \leq E(c \circ a, c \circ b), \ \forall a, b, c \in G.$$

Since the inequalities are valid for all $a, b, c \in G$, directly imply that

$$E(a \circ c, b \circ c) \leq E((a \circ c) \circ d, (b \circ c) \circ d)$$

and

$$E(c \circ a, c \circ b) \leq E(d \circ (c \circ a), d \circ (c \circ b)), \ \forall a, b, c, d \in G.$$

Choosing $d = c^{-1}$ we get

$$E(a \circ c, b \circ c) \leq E((a \circ c) \circ c^{-1}, (b \circ c) \circ c^{-1}) = E(a, b)$$

and

$$E(c \circ a, c \circ b) \leq E(c^{-1} \circ (c \circ a), c^{-1} \circ (c \circ c)) = E(a, b).$$

Therefore we have got

$$E(a, b) \leq E(a \circ c, b \circ c) \leq E(a, b) \text{ and}$$
$$E(a, b) \leq E(c \circ a, c \circ b) \leq E(a, b).$$

$\Leftarrow$) Trivial.

Proposition 12.15. *Let T-indistinguishability operator E on G. A binary operation $\circ$ on G is an extensional function with respect to $G(E, E)$ on $G \times G$ and E on G if and only if E is regular with respect to $\circ$.*

Proof

$\Rightarrow$) If $\circ$ is an extensional function with respect to $G(E, E)$ on $X \times X$ and E on X, then

$$G(E, E)((a, c), (b, d)) = T(E(a, b), E(c, d)) \leq E(a \circ c, b \circ d).$$

From this,

$$E(a, b) = T(E(a, b), E(c, c)) \leq E(a \circ c, b \circ c) \text{ and}$$
$$E(a, b) = T(E(c, c), E(a, b)) \leq E(c \circ a, c \circ b).$$
$$E(a, b) \leq E(a \circ c, b \circ c) \wedge E(c \circ a, c \circ b) \ \forall a, b, c \in G. \text{ Therefore}$$
$$E(a, b) \leq \bigwedge_{c \in G} E(a \circ c, b \circ c) \wedge E(c \circ a, c \circ b).$$

$\Leftarrow$) If E is regular with respect to $\circ$, then

$$E(a, b) \leq E(a \circ c, b \circ c) \text{ and}$$
$$E(a, b) \leq E(c \circ a, c \circ b) \ \forall a, b, c \in G.$$

Hence

$$\begin{aligned} G(E, E)((a, c), (b, d)) &= T(E(a, b), E(c, d)) \\ &\leq T(E(a \circ c, b \circ c), E(b \circ c, b \circ d) \\ &\leq E(a \circ c, b \circ d)). \end{aligned}$$

Corollary 12.16. *Let $(G, \circ)$ be a group and E a T-indistinguishability operator on G. $\circ$ is an extensional function with respect to $G(E, E)$ on $G \times G$ and E on G if and only if E is invariant under translations with respect to $\circ$.*

Proof. It is a consequence of Proposition 12.14 and Proposition 12.15.

Definition 12.17. *Let μ be a T-fuzzy subgroup of $(G, \circ)$ with $\mu(e) = 1$ where e is the identity element of G. The fuzzy relation $E_{(\mu)}$ on G defined by*

$$E_{(\mu)}(a, b) = \mu(a \circ b^{-1}) \ \forall a, b \in G$$

is the T-indistinguishability operator associated to μ.

Proposition 12.18. *$E_{(\mu)}$ on G is a T-indistinguishability operator on G.*

Proof

a) Reflexivity. Given $a \in G$, $E_{(\mu)}(a, a) = \mu(a \circ a^{-1}) = \mu(e) = 1$.
b) Symmetry.

$$\begin{aligned} E_{(\mu)}(a, b) &= \mu(a \circ b^{-1}) \\ &= \mu((a \circ b^{-1})^{-1}) \\ &= \mu(b \circ a^{-1}) = E_{(\mu)}(b, a). \end{aligned}$$

c) Transitivity.

$$\begin{aligned} T(E_{(\mu)}(a, b), E_{\mu}(b, c)) &= T(\mu(a \circ b^{-1}), \mu(b \circ c^{-1})) \\ &\leq \mu(a \circ b^{-1} \circ b \circ c^{-1}) \\ &= \mu(a \circ c^{-1}) = E_{(\mu)}(a, c). \end{aligned}$$

Lemma 12.19. *$E_{(\mu)}(a, b) = E_{(\mu)}(e, a \circ b^{-1}) \ \forall a, b \in G$.*

Proof. Trivial.

Reciprocally,

Proposition 12.20. *Let E be a T-indistinguishability operator on a group $(G, \circ)$ with identity element e such that for all $a, b \in G$ $E(a, b) = E(e, a \circ b^{-1})$. Then the column μ_e of E is a T-fuzzy subgroup of G with $\mu_e(e) = 1$.*

Proof

a)

$$\begin{aligned} T(\mu_e(a), \mu_e(b)) &= T(E(e, a), E(e, b)) \\ &\leq E(a, b) = E(e, a \circ b^{-1}) \\ &= \mu_e(a \circ b^{-1}). \end{aligned}$$

b) $\mu_e(e) = E(e, e) = 1$.

Corollary 12.21. *Let $(G, \circ)$ be a group with identity element e. There exists a bijection between the set of T-fuzzy subgroups of G and the set of T-indistinguishability operators on G satisfying Lemma 12.19 mapping every T-fuzzy subgroup μ of G into its associated T-indistinguishability operator $E_{(\mu)}$.*

The following definition fuzzifies the concept of normal subgroup. It will be one of the cornerstones of the chapter.

Definition 12.22. *A T-fuzzy subgroup μ of a group $(G, \circ)$ with identity element e is called a normal T-fuzzy subgroup if and only if $\mu(e) = 1$ and $\mu(a \circ b) \leq \mu(b \circ a)$ $\forall a, b \in G$.*

NB. Clearly, if μ is a normal T-fuzzy subgroup, then by symmetry $\mu(a \circ b) = \mu(b \circ a)$ $\forall a, b \in G$.

Proposition 12.23. *The core N of a normal T-fuzzy subgroup μ of a group $(G, \circ)$ is a normal subgroup of G.*

Proof

a) Proposition 12.4 assures that N is a subgroup of G.
b) $\forall a, b \in G$ if $a \circ b \in N$, then $b \circ a \in N$. Therefore N is a normal subgroup of G.

Proposition 12.24. *Let $(G, \circ)$ be a group with identity element e and μ a normal T-fuzzy subgroup of G with $\mu(e) = 1$. The associated T-indistinguishability operator $E_{(\mu)}$ of μ is invariant under translations.*

Proof. Let a, b, k be elements of G.

$$\begin{aligned} E_{(\mu)}(a, b) &= \mu(a \circ b^{-1}) = \mu(a \circ k \circ k^{-1} \circ b^{-1}) = E_{(\mu)}(a \circ k, b \circ k). \\ E_{(\mu)}(a, b) &= \mu(a \circ b^{-1}) = \mu(b^{-1} \circ a) = \mu(b^{-1} \circ k^{-1} \circ k \circ a) \\ &= \mu(k \circ a \circ b^{-1} \circ k^{-1}) = E_{(\mu)}(k \circ a, k \circ b). \end{aligned}$$

Reciprocally,

Proposition 12.25. *Let $(G, \circ)$ be a group with identity element e, μ a T-fuzzy subgroup of G with $\mu(e) = 1$ and $E_{(\mu)}$ its associated T-indistinguishability operator. If $E_{(\mu)}$ is invariant under translations, then μ is a normal T-fuzzy subgroup of G.*

Proof

$$\begin{aligned} \forall a, b \in G \ \mu(a \circ b) &= E_{(\mu)}(e, a \circ b^{-1}) = E_{(\mu)}(b, a) = \\ &= E_{(\mu)}(b \circ a^{-1}, e) \\ &= E_{(\mu)}(e, b \circ a^{-1}) = \mu(b \circ a). \end{aligned}$$

Corollary 12.26. *Let $G, \circ)$ be a group. There is a bijection between the normal fuzzy subgroups of G and the set of T-indistinguishability operators on G invariant under translations with respect to $\circ$.*

The following theorem links normality of a T-fuzzy subgroup μ with compatibility with respect to its associated T-indistinguishability operator $E_{(\mu)}$.

Theorem 12.27. *Let $(G, \circ)$ be a group with identity e, μ a T-fuzzy subgroup of G with $\mu(e) = 1$ and $E_{(\mu)}$ its associated T-indistinguishability operator. Then, $\circ$ is extensional with respect to $G(E_{(\mu)}, E_{(\mu)})$ and $E_{(\mu)}$ if and only if μ is normal.*

Proof. This is an immediate consequence of Proposition 12.24, Proposition 12.25 and Corollary 12.16.

12.3 Normal T-Fuzzy Subgroups and T-Vague Groups

The results of this section will allow us to interpret a T-vague group as the quotient of a group modulo a normal T-fuzzy subgroup.

Proposition 12.28. *Let E be a regular T-indistinguishability operator G with respect to a binary operation $\circ$ on G.*

a) A fuzzy relation $\tilde{\circ} : G \times G \times G \to [0, 1]$ satisfying the properties

$$\tilde{\circ}(x, y, x \circ y) = 1$$

and

$$\tilde{\circ}(x, y, z) \leq E(x \circ y, z) \text{ for all } x, y, z \in G,$$

is a T-vague binary operation on G with respect to $G(E, E)$ on $G \times G$ and E on X.

b) Furthermore, if $(G, \circ)$ is a semigroup, then $(G, \tilde{\circ})$ is a T-vague semigroup (i.e. it satisfies the associative property of Definition 12.7).

Proof. a) Since E is regular with respect to $\circ$, $\circ$ is an extensional function with respect to $G(E, E)$ and E thanks to Proposition 12.15. Thus, from Proposition 10.5 $\tilde{\circ}$ is a T-vague binary operation.

b) Associativity. Let $(G, \circ)$ be a semigroup. Using the associativity of $\circ$ and the regularity of E,

$$\begin{aligned}
&T(\tilde{\circ}(b, c, d), \tilde{\circ}(a, d, m), \tilde{\circ}(a, b, q), \tilde{\circ}(q, c, w)) \\
&\leq T(E(b \circ c, d), E(a \circ d, m), E(a \circ b, q), E(q \circ c, w)) \\
&\leq T(E(a \circ (b \circ c) \circ a \circ d), E(a \circ d, m), E((a \circ b) \circ c, q \circ c), E(q \circ c, w)) \\
&\leq T(E(a \circ (b \circ c) \circ m), E((a \circ b) \circ c, w)) \\
&\leq E(m, w) \; \forall a, b, c, d, m, q, w \in G.
\end{aligned}$$

Proposition 12.29. *Let $(G, \circ)$ be a group and E a T-indistinguishability operator on G invariant under translations with respect to $\circ$. Then for the fuzzy relation $\tilde{\circ} : G \times G \times G \to [0,1]$ defined by $\tilde{\circ}(x,y,z) = E(x \circ y, z)$ for all $x, y, z \in G$, $(G, \tilde{\circ})$ is a T-vague group with respect to E.*

Proof

a) Associativity. Proposition 12.14 and Proposition 12.28 directly give that $(G, \tilde{\circ})$ is a T-vague semigroup with respect to E.
b) Identity. Let e be the identity element of $(G, \circ)$. $\forall a \in G$

$$\tilde{\circ}(a, e, a) = E(a \circ e, a) = E(a, a) = 1.$$

Also

$$\tilde{\circ}(e, a, e) = E(e \circ a, a) = 1.$$

c) Inverse. Let a^{-1} be the inverse element of a in $(G, \circ)$.

$$\tilde{\circ}(a^{-1}, a, e) = E(a^{-1} \circ a, e) = E(e, e) = 1$$

and

$$\tilde{\circ}(a, a^{-1}, e) = E(a \circ a^{-1}, e) = E(e, e) = 1.$$

Corollary 12.30. *Let $(G, \circ)$ be a group with identity element e, μ a normal T-fuzzy subgroup of $(G, \circ)$ with $\mu(e) = 1$ and $E_{(\mu)}$ its associated T-indistinguishability operator on G. If $\tilde{\circ} : G \times G \times G \to [0,1]$ is defined for all $a, b, c \in G$ by $\tilde{\circ}(a, b, c) = \mu(a \circ b \circ c^{-1}) = E_{(\mu)}(a \circ b, c)$, then $(G, \tilde{\circ})$ is a T-vague group.*

Proof. $E_{(\mu)}$ is invariant under translations, since μ is a normal fuzzy subgroup. Then the result is a consequence of Proposition 12.29.

Proposition 12.31. *Let $(G, \tilde{\circ})$ be a T-vague group with respect to E and $\circ \in ORD(\tilde{\circ})$. Then,*

a) $\tilde{\circ}(x, y, z) = E(x \circ y, z)$ $\forall x, y, z \in G$.
b) $\circ$ is extensional with respect to $G(E, E)$ and E.
c) E is invariant under translations with respect to $\circ$.

Proof

a) is an immediate consequence of Proposition 10.10.2.
b) is also obvious from Proposition 10.7.
c) follows from 2. and Corollary 12.16.

Lemma 12.32. *Let $(G, \tilde{\circ})$ be a T-vague group with respect to E and $\circ \in ORD(\tilde{\circ})$. Then $E(a, b) = E(e, a \circ b^{-1})$ for each identity e of $(G, \tilde{\circ})$ and for each $a, b \in G$.*

Proof

$$\begin{aligned}E(a,b) &= T(\tilde{\circ}(a,b^{-1},a\circ b^{-1}),\tilde{\circ}(b,b^{-1},e),E(a,b),E(b^{-1},b^{-1}))\\ &\le E(e,a\circ b^{-1}).\\ E(e,a\circ b^{-1}) &= T(\tilde{\circ}(e,b,b),\tilde{\circ}(a\circ b^{-1},b,a),E(e,a\circ b^{-1}),E(b,b))\\ &\le E(a,b).\end{aligned}$$

At this point, given a group $(G,\circ)$ we have bijective maps between their T-vague groups, their T-fuzzy normal subgroups and their T-indistinguishability operators invariant under translations and a simple way to generate two of them knowing the other one: For instance, $\tilde{\circ}(a,b,c) = E(a\circ b,c) = \mu(a\circ b\circ c^{-1})$.

In [34], [35] homomorphisms between T-vague groups have been studied. In this Section we will use them to show that T-vague groups are in fact the fuzzy generalization of quotient groups and that if f is a homomorphism from $(G,\circ)$ to $(G,\tilde{\circ})$, then the kernel of f is precisely the normal fuzzy subgroup associated to $(G,\tilde{\circ})$.

Definition 12.33. *Let $(G,\tilde{\circ})$ and $(H,\tilde{\star})$ be two T-vague groups with respect to the T-indistinguishability operators E and F respectively. A map $f:G\to H$ is a homomorphism from G onto H if and only if*

$$\tilde{\circ}(a,b,c)\le\tilde{\star}(f(a),f(b),f(c))\ \forall a,b,c\in G.$$

Proposition 12.34. *Let $(G,\tilde{\circ})$ and $(G,\tilde{\star})$ be two T-vague groups with respect to the T-indistinguishability operators E and F respectively such that $E\le F$. Then the identity map* id $: G\to G$ *is a homomorphism from $(X,\tilde{\circ})$ onto $(X,\tilde{\star})$.*

Proof. It is an easy consequence of Proposition 12.31.

It is clear that a crisp group $(G,\circ)$ is a T-vague group defining $\circ(a,b,c)=1$ if $a\circ b=c$ and 0 otherwise, and considering the crisp equality as the T-indistinguishability operator.

Corollary 12.35. *Let $(G,\tilde{\circ})$ be a T-vague group with respect to a T-indistinguishability operator E and $\circ\in ORD(\tilde{\circ})$. Then the identity map* id $: G\to G$ *is a homomorphism from $(G,\circ)$ onto $(G,\tilde{\circ})$.*

Definition 12.36. *Let $f:G\to H$ be a homomorphism from $(G,\tilde{\circ})$ onto $(H,\tilde{\star})$. The kernel of f is the fuzzy subset μ of G defined by $\mu(a) = E(f(a),e)\ \forall a\in G$ where e is an identity element of $(H,\tilde{\star})$.*

Proposition 12.37. *Let $(G,\tilde{\circ})$ be a T-vague group with respect to a T-indistinguishability operator E and $\circ\in ORD(\tilde{\circ})$. The kernel of the identity map* id $: G\to G$ *is the normal T-fuzzy subgroup of G associated to $(G,\tilde{\circ})$.*

12.4 The Vague Real Line

In fuzzy control and other fuzzy systems it is usual to model the values of the linguistic variables by fuzzy numbers. Fuzzy numbers can be thought as equivalence classes of a indistinguishability operator in the sense that fuzzy numbers are needed because the equality used in the real line is not crisp due to lack of accuracy and vagueness. In many applications, fuzzy numbers of the same shape are used so that they only differ by a translation which means that the indistinguishability operator is invariant under translations. For this reason, it is interesting to determine when a T-indistinguishability operator on the real line is invariant under translations and has fuzzy numbers as their columns. In this section these relations will be characterized for the minimum and continuous Archimedean t-norms. As a corollary, fuzzy numbers that are also normal fuzzy subgroups of $(\mathbb{R}, +)$ and their corresponding T-vague groups will be characterized in these cases.

Let us remember that a fuzzy number is a map $\mu_a : \mathbb{R} \to [0,1]$ such that there exists $a \in \mathbb{R}$ with $\mu_a(a) = 1$ and non decreasing in $(-\infty, a)$ and non increasing in $(a, +\infty)$ (Definition 6.10).

Definition 12.38. *A T-indistinguishability operator on $\mathbb{R}$ is admissible if and only if its columns or singletons are fuzzy numbers.*

Lemma 12.39. *Let E be an T-indistinguishability operator on $(\mathbb{R}, +)$ invariant under translations. Then for every $a \in \mathbb{R}$ and $n \in \mathbb{N}$*

$$T(E(0,a), E(0,na)) \leq E(0,(n+1)a).$$

Proof. Trivial, since the column of E corresponding to 0 is a T-fuzzy subgroup.

Theorem 12.40. *Let T be the minimum t-norm. E is a T-indistinguishability operator on $(\mathbb{R}, +)$ admissible and invariant under translations if and only if there exists $k \in [0,1]$ such that*

$$E(a,b) = \begin{cases} 1 \text{ if } a = b \\ k \text{ otherwise.} \end{cases}$$

Proof. Let $a > 0$. From Lemma 12.39

$$\min(E(0,a), E(0,na)) \leq E(0,(n+1)a) \quad \forall n \in \mathbb{N}$$

and since E is admissible, $\min(E(0,a), E(0,na)) = E(0,na)$ so that

$$E(0,na) \leq E(0,(n+1)a).$$

But since E is admissible, $E(0,na) \geq E(0,(n+1)a)$.

Therefore $E(0,na) = E(0,(n+1)a)$ $\forall a > 0$, $\forall n \in N$. Due to the monotonicity of $E(0,\cdot)$ in $(0,+\infty)$, $E(0,\cdot)$ must be constant in this interval.

A similar argument works with $a < 0$ for $(-\infty, 0)$.

Now the result follows applying the invariance under translations of E.

Corollary 12.41. *A fuzzy number μ is a normal* min*-fuzzy subgroup of $(\mathbb{R}, +)$ if and only if there exists $k \in \mathbb{R}$ such that*

$$\mu(b) = \begin{cases} 1 \text{ if } b = 0 \\ k \text{ otherwise.} \end{cases}$$

Corollary 12.42. *The T-vague group $(\mathbb{R}, \tilde{+})$ with respect to the* min*-indistinguishability operator E defined by $E(a,b) = k$ if $a \neq b$ is*

$$\tilde{+}(a,b,c) = \begin{cases} 1 \text{ if } a + b = c \\ k \text{ otherwise.} \end{cases}$$

Theorem 12.43. *Let T be a continuous Archimedean t-norm with generator t. E is an admissible and invariant under translations T-indistinguishability operator on $(\mathbb{R}, +)$ if and only if there exists a map $F : \mathbb{R} \to \mathbb{R}$ which is non decreasing and subadditive in $\mathbb{R}^+$ with $F(0) = 0$ and $F(a) = F(-a)$ $\forall a \in R$ such that $E(a,b) = t^{[-1]} \circ F(b - a)$.*

Proof

$\Rightarrow$) If E is invariant under translations, then

$$T(E(0,a), E(0,c)) \leq E(0, a + c)$$

since the column of E corresponding to 0 is a T fuzzy subgroup. This inequality can be rewritten as

$$t^{[-1]}(t(E(0,a)) + t(E(0,c))) \leq E(0, a + c)$$

or

$$t(E(0,a)) + t(E(0,c)) \geq t(E(0, a + c)).$$

If F is defined by $F(x) = t(E(0,x))$, then $F(a) + F(b) \geq F(a + b)$ and $F(0) = 0$, $F(x) = F(-x)$ and is non decreasing in $(0, +\infty)$.

$E(0,a) = t^{[-1]}(F(a))$ and $E(a,b) = E(0, b - a) = t^{[-1]}(F(b - a))$.

$\Leftarrow$). Trivial.

Corollary 12.44. *Let T be a continuous Archimedean t-norm with generator t. A fuzzy number μ_a is a normal T-fuzzy subgroup of $(\mathbb{R}, +)$ if and only if there exists a map $F : \mathbb{R} \to \mathbb{R}$ which is non decreasing and subadditive in $\mathbb{R}^+$ with $F(0) = 0$ and $F(x) = F(-x)$ $\forall x \in R$ such that $\mu(a) = t^{[-1]} \circ F(a)$.*

Corollary 12.45. *Let T be a continuous Archimedean t-norm with generator t. and $\tilde{+} : \mathbb{R} \times \mathbb{R} \times \mathbb{R} \to [0,1]$. $(\mathbb{R}, \tilde{+})$ is a T-vague group with respect to an admissible T-indistinguishability operator E if and only if there exists a map $F : \mathbb{R} \to \mathbb{R}$ which is non decreasing and subadditive in $\mathbb{R}^+$ with $F(0) = 0$ and $F(a) = F(-a)$ $\forall a \in \mathbb{R}$ such that $\tilde{+}(a,b,c) = t^{[-1]} \circ F(a + b - c)$.*

This section ends with two examples illustrating the results of the chapter. The first one specifies the last results to triangular numbers while the second one is a first attempt to define the property of being a vague multiple of an integer.

Example 12.46. Let μ be a symmetric triangular number centered in 0. (i.e. $\mu = [-k, 0, k]$ for some $k \in \mathbb{R}$). Taking $F(x) = |x/k|$ and $t = 1 - x$ in Corollary 12.44, μ is a normal T-fuzzy subgroup of $(\mathbb{R}, +)$, where T stands for the Łukasiewicz t-norm.

The associated T-indistinguishability operator $E_{(\mu)}$ defined by $E_{(\mu)}(a, b) = \mu(a - b)$ is separable and invariant under translations and therefore all their columns (i.e. all lateral classes of μ) are triangular numbers of the same shape.

The T-vague group associated to μ is $(\mathbb{R}, \tilde{+})$ where

$$\tilde{+}(a, b, c) = \mu(a + b - c) = E(e, a + b - c).$$

This means that the vague sum of two numbers a and b is the symmetric triangular number translated of μ and centered in $a + b$.

The following example considers the possibility of being a vague multiple of an integer.

Example 12.47. Given an integer number a and a symmetric triangular number μ centered at 0 and with its core reduced to $\{0\}$, let us consider for each multiple na of a the fuzzy number μ_{na} centered at na and translated of μ (i.e. $\mu_{na}(b) = \mu(na - b)$). Let us define the fuzzy set $\mu_{\overset{\circ}{a}}$ by

$$\mu_{\overset{\circ}{a}}(b) = \sup_{n \in \mathbb{Z}} \mu_{na}(b) \ \forall b \in \mathbb{R}.$$

$\mu_{\overset{\circ}{a}}$ is a normal T-fuzzy subgroup of $(\mathbb{R}, +)$ (T the Łukasiewicz t-norm) and $\mu_{\overset{\circ}{a}}(b)$ is the degree of b being multiple of a.

The core of $\mu_{\overset{\circ}{a}}$ is the subgroup (a) of multiples of a.

The associated T-vague group of $\mu_{\overset{\circ}{a}}$ can be thought as the fuzzification of the quotient group $\mathbb{Z}/(a)$ of the integer numbers modulo a.

13

Finitely Valued Indistinguishability Operators

The literature contains examples of indistinguishability operators valued in more general structures than the unit interval endorsed by a t-norm. In [60] [58], for example, indistinguishability operators are studied under category theory. In [23], the unit interval was generalized to GL-monoids. These generalizations are very useful, because they simplify the study of some specific cases. For example, many of the results found in this chapter about finitely valued indistinguishability operators can be proved in exactly the same way as in the unit interval case because both are GL-monoids.

Indistinguishability operators have also been studied in specific frameworks. For instance, in [115] indistinguishability operators valued on probabilistic metric spaces were studied and a Representation Theorem similar to 2.54 was proved for them.

Interval-valued indistinguishability operators have also been studied ([51]). The cited study introduced a new concept of transitivity, called weak transitivity, that coincides with classical T-transitivity for $[0,1]$ valued indistinguishability operators but does not in general. (A fuzzy interval-valued relation R on X is weak transitive if and only if for every $x, y \in X$ there does not exist $z \in X$ such that $\mathcal{T}(R(x,z), R(z,y)) > R(x,y)$, where $\mathcal{T}$ is a generalized t-norm defined on the set of intervals of $[0,1]$). This is an example of how a concept can be developed in specific domains.

Finite-valued indistinguishability operators are another very interesting kind of operators. This is because in most situations, the properties of objects are divided according to a finite set of linguistic labels such as *very small, small, medium, big* and *very big*. A finite valued t-norm T makes it possible to calculate with these labels directly and obtain, for example, the result of $T(small, very\ big)$ as another linguistic label. This chapter is devoted to the study of such indistinguishability operators. Some preliminaries on finite-valued t-norms are provided in Section 13.1 in order to make this discussion self-contained. Section 13.2 presents some properties of finite-valued indistinguishability operators. In particular, the Representation Theorem is generalized to these operators. Section 13.3 is about additive generators of

J. Recasens: Indistinguishability Operator, STUDFUZZ 260, pp. 217–229.
springerlink.com

finite-valued t-norms. Unlike t-norms valued in $[0,1]$, most finite-valued t-norms have additive generators. A new pseudoinverse is defined in order to generate their residuation. The results are applied in Section 13.4 to find the dimension and a basis of finite-valued indistinguishability operators by solving some systems of Diophantine inequalities. The results can be applied to find the dimension and a basis of usual $[0,1]$-valued T-indistinguishability operators, and new algorithms for doing this are provided (Section 13.5).

Finite-valued t-norms

In fuzzy logic, the logical conjunction is modelled by a t-norm. In this way infinitely valued logics are obtained in which the truth degree of a proposition is a number between 0 and 1. In fuzzy systems, t-norms are also used to model the intersection of fuzzy subsets that are valued in the unit interval.

In many cases, assigning an exact and precise value between 0 and 1 is not realistic because, due to linguistic vagueness or lack of precision in the data this assignment is necessarily imprecise. It would be more reasonable in these cases to consider only a totally ordered finite chain (that can be identified with a finite subset of [0,1]) in order to valuate the fuzzy concepts.

The study of operators defined on a finite chain L is of great interest, especially because reasoning is usually done by using linguistic terms or labels that are totally ordered. For instance, the size of an object can be granularized into *very small, small, average, big, very big.* If an operator T is defined on this set, then we will be able to combine these labels in order to obtain for example $T(average, very\ big)$. The calculations are simplified greatly by addressing the problem of combining labels in this way, since there is no need to assign numerical values to them or to identify them with an interval or with a fuzzy subset.

Finite chains are also useful in cases in which the values are discrete by nature or by discretization. On a customer-satisfaction survey respondents may be asked to describe their satisfaction with a service using natural number from 0 to 5 or labels ranging from *not at all satisfied* to *very satisfied.*

In this line, various authors have translated t-norms and t-conorms to finite chains ([92], [93]) and have obtained interesting theoretical results.

Finite-valued Indistinguishability Operators

In almost all situations, human beings categorize or granularize the properties or features of objects into a finite set L of linguistic labels that can be linearly ordered. These properties can be evaluated on L in a natural way and, consequently, the fuzzy subsets of the universe of discourse are also valued on L.

Likewise, the degree of similarity, equivalence or indistinguishability between two objects is not a numerical value between 0 and 1, but rather an element of L that can be interpreted as *rather, very much,* etc.

Indistinguishability operators valued in finite chains seem to be very interesting tools that will allow us to study the similarity between objects while taking into account the granularity generated by L and obtain an interpretation of the calculation on the chain.

13.1 Preliminaries

This section contains some definitions and results on finite valued t-norms that will be needed on this chapter.

Let $L =$ be a finite totally ordered set with minimum e and maximum u.

Definition 13.1. *A binary operation $T : L \times L \to L$ is a t-norm if and only if for all $x, y, z \in L$*

1. $T(x, y) = T(y, x)$
2. $T(T(x, y), z) = T(x, T(y, z))$
3. $T(x, y) \leq T(x, z)$ *whenever* $y \leq z$
4. $T(x, u) = x$

The set of t-norms on a finite chain depends only on its cardinality. For this reason we will only consider the chains $L = \{0, 1, ..., n\}$ and $L' = \{0 = \frac{0}{n}, \frac{1}{n}, \frac{2}{n}, ..., \frac{n}{n} = 1\}$.

Example 13.2

1. The minimum t-norm on L is defined by

$$T(i, j) = \min(i, j) \ \forall i, j \in L.$$

2. The Łukasiewicz t-norm on L is defined by

$$T(i, j) = \max(i + j - n, 0) \ \forall i, j \in L.$$

Smooth t-norms on finite chains are the equivalent of continuous ones defined on $[0, 1]$.

Definition 13.3. [93]

- *A map $f : L \to L$ is smooth if and only if*

$$0 \leq f(i + 1) - f(i) \leq 1 \text{ for all } i \in L.$$

- *A map $f : L \times L \to L$ is smooth if and only if it is smooth with respect to both variables.*

Definition 13.4. [93] *A t-norm T on L is divisible if and only if for all $i, j \in L$ with $i \leq j$ there exists $k \in L$ such that*

$$i = T(j, k).$$

Smoothness and divisibility are equivalent concepts for finite valued t-norms.

Proposition 13.5. [93] *A t-norm on L is smooth if and only if it is divisible.*

The next proposition characterizes all smooth t-norms on L as particular ordinal sums of copies of the t-norm of Łukasiewicz.

Proposition 13.6. [93] *A t-norm T on L is smooth if and only if there exists $J = \{0 = i_0 < i_1 < ... < i_m = n\} \subseteq L$ such that*

$$T(i,j) = \begin{cases} \max(i_k, i+j-i_k) & \text{if } i,j \in [i_k, i_{k+1}] \text{ for some } i_k \in J \\ \min(i,j) & \text{otherwise.} \end{cases}$$

T is said to be an ordinal sum and will be represented by $T =< 0 = i_0, i_1, ...i_m = n >$.

13.2 Finitely Valued Indistinguishability Operators

The Representation Theorem of indistinguishability operators was first proved by Ovchinnikov for the Product t-norm [105], then it was generalized to continuous t-norms by Valverde in [139] and in [23] it is noticed that it is also true for GL-monoids. Since finite valued t-norms are such monoids, the Representation Theorem also applies to finitely valued indistinguishability operators.

This section adapts the basic definitions and properties of indistinguishability operators to the finite valued case. In particular, the Representation Theorem and the idea of extensionality are reformulated. Also the concepts of dimension and basis of an indistinguishability operator are considered and the characterization of the set of extensional fuzzy subsets with respect to an indistinguishability operator is adapted to the context of finite valued t-norms.

Definition 13.7. *Let T be a t-norm on L. Its residuation $\overrightarrow{T}$ is defined by*

$$\overrightarrow{T}(i|j) = \max\{k \in L \mid T(i,k) \leq j\}.$$

Example 13.8

1. If T is the Łukasiewicz t-norm on L, then $\overrightarrow{T}(i|j) = \max(0, n-i+j)$ for all $i,j \in L$.
2. If T is the Minimum t-norm on L, then $\overrightarrow{T}(i|j) = \begin{cases} j & \text{if } i > j \\ n & \text{otherwise.} \end{cases}$

Proposition 13.9. *Let $T =< 0 = i_0, i_1, ...i_m = n >$ be a smooth t-norm on L. Its residuation $\overrightarrow{T}$ is*

$$\overrightarrow{T}(i|j) = \begin{cases} n & \text{if } i \leq j \\ \max(i_k, i_{k+1} - i + j) & \text{if } i,j \in [i_k, i_{k+1}] \text{ for some } i_k \in J \text{ and } i > j \\ j & \text{otherwise.} \end{cases}$$

Definition 13.10. *The biresiduation E_T associated to a given t-norm T on L is defined by*

$$E_T(i,j) = T(\overrightarrow{T}(i|j), \overrightarrow{T}(j|i)) = \min(\overrightarrow{T}(i|j), \overrightarrow{T}(j|i)).$$

Example 13.11

1. If T is the Łukasiewicz t-norm on L, then $E_T(i,j) = n - |i-j|$ for all $i,j \in L$.
2. If T is the Minimum t-norm, then $E_T(i,j) = \begin{cases} \min(i,j) & \text{if } i \neq j \\ n & \text{otherwise.} \end{cases}$

Proposition 13.12. *Let $T =< 0 = i_0, i_1, ... i_m = n >$ be a smooth t-norm on L. Its biresiduation E_T is*

$$E_T(i,j) = \begin{cases} n & \text{if } i = j \\ i_{k+1} - |i-j| & \text{if } i,j \in [i_k, i_{k+1}] \text{ for some } i_k \in J \text{ and } i \neq j \\ \min(i,j) & \text{otherwise.} \end{cases}$$

$\overrightarrow{T}$ and E_T are special kind of T-preorders and T-indistinguishability operators.

Definition 13.13. *Let T be a t-norm on L. A T-preorder P on a set X is a fuzzy relation $P : X \times X \to L$ satisfying for all $x, y, z \in X$*

1. *$P(x,x) = n$ (Reflexivity)*
2. *$T(P(x,y), P(y,z)) \leq P(x,z)$ (T-transitivity).*

Definition 13.14. *Let T be a t-norm on L. A T-indistinguishability operator E on a set X is a fuzzy relation $E : X \times X \to L$ satisfying for all $x, y, z \in X$*

1. *$E(x,x) = n$ (Reflexivity)*
2. *$E(x,y) = E(y,x)$ (Symmetry)*
3. *$T(E(x,y), E(y,z)) \leq E(x,z)$ (T-transitivity).*

Proposition 13.15. *The residuation $\overrightarrow{T}$ of a t-norm T on L is a T-preorder on L.*

Proposition 13.16. *The biresiduation E_T of a t-norm T on L is a T-indistinguishability operator on L.*

Lemma 13.17. *Let T be a t-norm on L and μ an L-fuzzy subset of X (i.e., $\mu : x \to L$). The fuzzy relation E_μ on X defined for all $x, y \in X$ by*

$$E_\mu(x,y) = E_T(\mu(x), \mu(y))$$

is a T-indistinguishability operator.

Theorem 13.18. *Representation Theorem for T-indistinguishability operators. Let R be a fuzzy relation on a set X and T a t-norm on L. R is a T-indistinguishability operator if and only if there exists a family $(\mu_i)_{i\in I}$ of L-fuzzy subsets of X such that for all $x, y \in X$*

$$R(x,y) = \inf_{i\in I} E_{\mu_i}(x,y).$$

$(\mu_i)_{i\in I}$ is called a generating family of R and an L-fuzzy subset that belongs to a generating family of R is called a generator of R.

A similar result holds for T-preorders.

As in the $[0,1]$ valued case, extensional L-fuzzy subsets with respect to a finitely valued T-indistinguishability operator E play a central role since they are the only observable sets taking E into account.

Extensional sets with respect to a T-indistinguishability operator coincide with their generators, as is stated in Proposition 13.20.

Definition 13.19. *Let T be a t-norm on L, E a T-indistinguishability operator on a set X and μ an L-fuzzy subset of X. μ is extensional with respect to E if and only if for all $x, y \in X$*

$$T(E(x,y), \mu(x)) \leq \mu(y).$$

H_E will denote the set of all extensional L-fuzzy subsets with respect to E.

It can be proved that an L-fuzzy subset is extensional with respect to a T-indistinguishability operator E if and only if it is a generator of E.

Proposition 13.20. *Let T be a t-norm on L, E a T-indistinguishability operator on a set X and μ an L-fuzzy subset of X. μ is extensional with respect to E if and only if $E_\mu \geq E$.*

In the same way as in the $[0,1]$ valued case, the concepts of generating family, dimension and basis can be defined.

Definition 13.21. *Let T be a t-norm on L and E a T-indistinguishability operator on X. The dimension of E is the minimum of the cardinalities of the generating families of E in the sense of the Representation Theorem. A generating family with this cardinality is called a basis of E.*

In Section 13.4 an algorithm to find dimensions and basis of T-indistinguishability operators for an additively generated t-norm T on L will be provided.

13.3 Additive Generators

Many of the t-norms on a finite chain L can be additively generated. In particular, it can be proved that all smooth t-norms on L - including the minimum t-norm and all ordinal sums - have an additive generator. A t-norm T on L with additive generator f can then be generated combining f with its pseudoinverse $f_+^{(-1)}$. A new pseudoinverse $f_-^{(-1)}$ will be useful to generate $\overrightarrow{T}$ and E_T. These representations will provide us in the next Section 13.4 with a technique to find the dimension and a basis of a finitely valued T-indistinguishability operator E as well as its set H_E of generators or extensional sets when T has an additive generator.

Definition 13.22. *Let $f : L \to [0, \infty)$ be a strictly decreasing function with $f(n) = 0$.*

- *The pseudo inverse $f_+^{(-1)} : [0, \infty) \to L$ of f is defined by*

$$f_+^{(-1)}(t) = \min\{i \in L; f(i) \leq t\} = \min f^{-1}([0, t]).$$

- *The pseudo inverse $f_-^{(-1)} : (-\infty, \infty) \to L$ of f is defined by*

$$f_-^{(-1)}(t) = \begin{cases} \max\{i \in L; f(i) \geq t\} = \max f^{-1}([t, n]) & \text{if } t \geq 0 \\ n & \text{otherwise.} \end{cases}$$

The first pseudo inverse $f_+^{(-1)}$ was first defined in [93]. $f_-^{(-1)}$ is a new pseudo inverse introduced here in order to generate the residuation and biresiduation of a t-norm on L.

Definition 13.23. [93] *Let T be a t-norm on L. T is generated by a strictly decreasing function $f : L \to [0, \infty)$ with $f(n) = 0$ if and only if*

$$T(i, j) = f_+^{(-1)}(f(i) + f(j)) \text{ for all } i, j \in L.$$

f is called an additive generator of T and we will write $T = \langle f \rangle$.

For an additive generator f we will indicate $f = (a_0, a_1, a_2, ..., a_n = 0)$ where $a_i = f(i), i \in L$.

Example 13.24

- An additive generator of the t-norm of Łukasiewicz on L is $(n, n-1, n-2, ..., 1, 0)$.
- An additive generator of the minimum t-norm on L is $(2^n - 1, 2^{n-1} - 1, 2^{n-2} - 2, ..., 7, 3, 1, 0)$.

The next results present some intersting properties of additive generators.

Proposition 13.25. [93] *Let $f = (a_0, a_1, a_2, ..., a_n = 0)$ and $g = (b_0, b_1, b_2, ..., b_n = 0)$ be strictly decreasing functions on L. Then $\langle f \rangle = \langle g \rangle$ if and only if for all $i, j, k \in L$ with $k \neq 0$,*

1. $a_i + a_j \geq a_0 \Rightarrow b_i + b_j \geq 0$
2. $a_k \leq a_i + a_j < a_{k-1} \Rightarrow b_k \leq b_i + b_j < b_{k-1}$.

Corollary 13.26. [93] *If* $f : L \to [0, \infty)$ *is a strictly decreasing function with* $f(n) = 0$ *and* $\lambda > 0$*, then* $\langle f \rangle = \langle \lambda f \rangle$.

Of course, the reciprocal of the corollary is not true.

Proposition 13.27. [93] *If* T *is a t-norm on* L *with additive generator, then we can find an additive generator* f *of* T *with Ran* $f \in \mathbb{Z}^+$.

Proposition 13.28. [93] *Al smooth t-norms on* L *have an additive generator.*

For additively generated t-norms we have representations for their residuations and biresiduations.

Proposition 13.29. *Let* T *be a t-norm on* L *with additive generator* f*. Then*

$$\overrightarrow{T}(i|j) = f_-^{(-1)}(f(j) - f(i)) \text{ for all } i, j \in L.$$

Proof. Given $i, j \in L$,

$$\begin{aligned}\overrightarrow{T}(i|j) &= \max\{k \in L \mid T(i,k) \leq j\} \\ &= \max\{k \in L \mid f_+^{(-1)}(f(i) + f(k)) \leq j\} \\ &= f_-^{(-1)}(f(j) - f(i)).\end{aligned}$$

Proposition 13.30. *Let* T *be a t-norm on* L *with additive generator* f*. Then*

$$E_T(i,j) = f_-^{(-1)}(|f(i) - f(j)|) \text{ for all } i, j \in L.$$

Proof

$$\begin{aligned}E_T(i,j) &= \min(\overrightarrow{T}(i|j), \overrightarrow{T}(j|i)) \\ &= \min(f_-^{(-1)}(f(j) - f(i)), f_-^{(-1)}(f(i) - f(j))) \\ &= f_-^{(-1)}(|f(i) - f(j)|).\end{aligned}$$

13.4 Dimension and Basis of an Indistinguishability Operator

In this section a method to calculate the dimension an a basis of a T-indistinguishability operator E on a finite set X when T, a t-norm on L, can be additively generated willbe given.

Also a characterization of the T-transitive closure of a reflexive and symmetric fuzzy relation will be provided.

Let μ be an L-fuzzy subset of a finite set $X = \{r_1, r_2, ..., r_s\}$ of cardinality s. We will write $\mu = (q_1, q_2, ..., q_s)$ when $\mu(r_i) = q_i, i = 1, 2, ..., s$.

An L-fuzzy subset μ of X is a generator of E if and only if $E_\mu(r_i, r_j) \geq E(r_i, r_j)$ for all $i, j = 1, 2, ..., s$. If T has f as an additive generator, then this condition can be written as

$$f_-^{(-1)}(|f(\mu(r_i)) - f(\mu(r_j))|) \geq E(r_i, r_j) \text{ for all } i, j = 1, 2, ..., s$$

or

$$|f(\mu(r_i)) - f(\mu(r_j))| \leq f(E(r_i, r_j)) \text{ for all } i, j = 1, 2, ..., s.$$

This is equivalent to

$$f(\mu(r_i)) - f(\mu(r_j)) \leq f(E(r_i, r_j)) \text{ for all } i, j = 1, 2, ..., s.$$

Proposition 13.31. *Let T be a t-norm on L with additive generator f and E a T-indistinguishability operator on a finite set X of cardinality s. An L-fuzzy subset $\mu = (x_1, x_2, ..., x_s)$ is a generator of E if and only if*

$$f(x_i) - f(x_j) \leq f(E(r_i, r_j)) \text{ for all } i, j = 1, 2, ..., s.$$

In other words, H_E is the subset of L^s of solutions of the last system of Diophantine inequalities.

Example 13.32. If T is the Łukasiewicz t-norm on L, then the last system of inequalities becomes

$$x_i - x_j \leq n - E(r_i, r_j) \text{ for all } i, j = 1, 2, ..., s.$$

Example 13.33. If T is the minimum t-norm on L, then the last system of inequalities becomes

$$2^{n-x_i} - 2^{n-x_j} \leq 2^{n-E(r_i,r_j)} - 1 \text{ for all } i, j = 1, 2, ..., s.$$

Example 13.34. The following fuzzy relation E on $X = \{r_1, r_2, r_3, r_4\}$ is a min-indistinguishability operator with $L = \{0, 1, 2\}$.

$$E = \begin{pmatrix} 2 & 1 & 0 & 0 \\ 1 & 2 & 0 & 0 \\ 0 & 0 & 2 & 1 \\ 0 & 0 & 1 & 2 \end{pmatrix}.$$

An L-fuzzy subset (x_1, x_2, x_3, x_4) of X is a generator of E if and only if it satisfies the following system of Diophantine inequations.

$$\begin{aligned} 2^{2-x_1} - 2^{2-x_1} &\leq 2^{2-1} - 1 = 1 \\ 2^{2-x_1} - 2^{2-x_3} &\leq \ 2^2 - 1 = 3 \\ 2^{2-x_1} - 2^{2-x_4} &\leq \quad 3 \\ 2^{2-x_2} - 2^{2-x_1} &\leq \quad 1 \end{aligned}$$

$$\begin{aligned}
2^{2-x_2} - 2^{2-x_3} &\leq 3 \\
2^{2-x_2} - 2^{2-x_4} &\leq 3 \\
2^{2-x_3} - 2^{2-x_1} &\leq 3 \\
2^{2-x_3} - 2^{2-x_2} &\leq 3 \\
2^{2-x_3} - 2^{2-x_4} &\leq 1 \\
2^{2-x_4} - 2^{2-x_1} &\leq 3 \\
2^{2-x_4} - 2^{2-x_2} &\leq 3 \\
2^{2-x_4} - 2^{2-x_3} &\leq 1
\end{aligned}$$

H_E has 26 elements:

$$\begin{aligned}
H_E = \quad & \{(2,2,2,2),(2,2,2,1),(2,2,1,2),(2,2,1,1),(2,2,0,0), \\
& (2,1,2,2),(2,1,2,1),(2,1,1,2),(2,1,1,1),(2,2,0,0), \\
& (2,1,0,0),(1,2,2,2),(1,2,2,1),(1,2,1,2),(1,2,1,1), \\
& (1,2,0,0),(1,1,2,2),(1,1,2,1),(1,1,1,2),(1,1,1,1), \\
& (1,1,0,0),(0,0,1,2),(0,0,2,1),(0,0,2,2),(0,0,1,1),(0,0,0,0)\}
\end{aligned}$$

E has dimension 2 and $\{(1,2,0,0),(0,0,1,2)\}$ is a basis of E.

The next result shows when two L-fuzzy subsets generate the same T-indistinguishability operator, T an L valued t-norm with additive generator. The proof is very similar to the proof of Theorem 4.23 in Chapter 4.

Proposition 13.35. *Let μ, ν be two L-fuzzy subsets of X and T a t-norm on L with additive generator f. $E_\mu = E_\nu$ if and only if $\forall x \in X$ one of the following conditions holds:*

1. $f(\mu(x)) = f(\nu(x)) + k_1$ *with* $\min\{f(0) - f(\nu(x)) \leq k_1 \geq \max\{-f(\nu(x)) | x \in X\}$
2. $f(\mu(x)) = -f(\nu(x)) + k_2$ *with* $\min\{f(0) + f(\nu(x)) \leq k_2 \geq \max\{f(\nu(x)) | x \in X\}$.

Proposition 13.36. *Let T be a t-norm on L with additive generator f and E a T-indistinguishability operator on X. Then $f \circ E$ is a pseudometric on X. Reciprocally, if m is a pseudometric on X, then $f_-^{(-1)} \circ m$ is a T-indistinguishability operator on X.*

E is a min*-indistinguishability operator if and only if $f \circ E$ is a pseudo ultrametric.*

This section ends with a characterization of the T-transitive closure (T a t-norm valued on L) of a reflexive and symmetric fuzzy relation using the set of its extensional L-fuzzy subsets or generators.

Proposition 13.37. *Let T be a t-norm on L, X a finite set and R a proximity relation on X valued on L. The transitive closure $\overline{R}$ of R is the T-indistinguishability operator on X valued on L having as extensional sets the set $H_{\overline{R}}$ of L-fuzzy subsets μ of X with $E_\mu \geq R$.*

Proof

$$\mu \in H_{\overline{R}} \text{ if and only if } E_\mu \geq \overline{R} \geq R.$$

Example 13.38. Let us consider the proximity relation R on $X = \{x_1, x_2, x_3\}$ valued on $L = \{0, 1, 2, 3\}$ with matrix

$$R = \begin{pmatrix} 3 & 2 & 0 \\ 2 & 3 & 2 \\ 0 & 2 & 3 \end{pmatrix}.$$

For the Łukasiewicz t-norm, an L-fuzzy subset (x_1, x_2, x_3) of X is in $H_{\overline{R}}$ if and only if it satisfies the following Diophantine system of inequations.

$$\begin{aligned} x_1 - x_2 &\leq 3 - R(x_1, x_2) = 3 - 2 = 1 \\ x_2 - x_1 &\leq 1 \\ x_1 - x_3 &\leq 3 \\ x_3 - x_1 &\leq 3 \\ x_2 - x_3 &\leq 1 \\ x_3 - x_2 &\leq 1. \end{aligned}$$

$$\begin{aligned} H_{\overline{R}} = \{&(0,0,0), (0,0,1), (0,1,0), (0,1,1), (0,1,2), (1,0,0), (1,0,1), \\ &(1,1,0), (1,1,1), (1,1,2), (1,2,1), (1,2,2), (1,2,3), (2,1,1), \\ &(2,1,2), (2,2,1), (2,2,2), (2,2,3), (2,3,2), (2,3,3), (3,2,1), \\ &(3,2,2), (3,2,3), (3,3,2), (3,3,3), (3,3,4), (3,4,3), (3,4,4), \\ &(4,3,2), (4,3,3), (4,3,4), (4,4,3), (4,4,4)\}. \end{aligned}$$

The T-transitive closure $\overline{R}$ of R is generated by $H_{\overline{R}}$ and its matrix is

$$\overline{R} = \begin{pmatrix} 4 & 3 & 2 \\ 3 & 4 & 3 \\ 2 & 3 & 4 \end{pmatrix}.$$

It has dimension 1 and $\{(0, 1, 2)\}$ is a basis of $\overline{R}$.

13.5 Approximation of Indistinguishability Operators Valued on [0,1] by Finitely Valued Ones

The problem of finding the dimension and a basis of a T-indistinguishability operator on a finite set X for a continuous t-norm T on [0,1] has been treated in Chapter 7.

For practical purposes, if X is finite we can assume that the entries of a T-indistinguishability operator are rational numbers, since in many cases the data come from inexact measurements and also because every T-indistinguishability operator can be approximated by another one with rational entries with as much accuracy as desired.

Having this in mind, the results obtained in the previous sections will be used to find the dimension and a basis of a T-indistinguishability operator as close as needed to a given one for T the minimum, the Łukasiewicz t-norm or any ordinal sum of finite copies of the Łukasiewicz t-norm with no segments of idempotent elements on the diagonal.

Definition 13.39. *Let R and S be two fuzzy relations on a finite set X. The distance $\|R - S\|$ between R and S is*

$$\|R - S\| = \max(|R(x, y) - S(x, y)| \; x, y \in X).$$

Lemma 13.40. *Let r be a positive real number and n a positive integer. There exists a rational number $q = \frac{a}{n}$ with $q \leq r$ and $r - q \leq \frac{1}{n}$.*

Proof. $0 \leq r - q \leq \frac{1}{n}$ is equivalent to $nr - 1 \leq a \leq nr$. Take a the natural number satisfying both inequalities.

Proposition 13.41. *Let T be an ordinal sum on $[0, 1]$ of a finite number of Łukasiewicz copies on the boxes $[a_{i-1}, a_i]^2$, $i = 1, 2, ..., t$ with $0 = a_0 < a_1 < a_2 < ... < a_t = 1$, $a_i = \frac{b_i}{n}$, $b_i \in Z$ for $i = 0, .., t$ and n a fixed integer. Let R a proximity relation on a finite set X its entries fractions with divisor n (i.e., $R(x, y) = \frac{s_{xy}}{n}$ with s_{xy} an integer for all $x, y \in X$). Then the entries of its T-transitive closure are fractions with divisor n as well.*

Proof. The T-transitive closure of R can be obtained using the $\sup -T$ product than involves only sums, subtractions, the maximum and the minimum.

Proposition 13.42. *Let T be an ordinal sum on $[0, 1]$ as in the preceding proposition, E a T-indistinguishability operator on a finite set X and $n \in \mathbb{N}$. There exists a T-indistinguishability operator E' on X smaller than or equal to E with all its entries fraction numbers with denominator n and such that $\|E - E'\| < \frac{1}{n}$.*

Proof. Let $a_1, a_2, ..., a_p$ be the entries of E different from 1. We can find $a'_1, a'_2, ..., a'_p$ rational numbers with denominator n such that $a_i - a'_i < \frac{1}{n}$ and $a'_i \leq a_i$ for all $i = 1, 2, ..., p$. Replacing the entries $a_1, a_2, ..., a_p$ of E by $a'_1, a'_2, ..., a'_p$ respectively, we obtain a new reflexive and symmetric fuzzy relation E'' with all its entries rational numbers with denominator n and satisfying $\|E - E''\| < \frac{1}{n}$. E'' may not be T-transitive. Its T-transitive closure E' is between E and E''; therefore $\|E - E'\| < \frac{1}{n}$. Moreover, thanks to Proposition 13.41, the entries of E' are rational numbers with denominator n.

This proposition allows us to calculate the dimension and a basis of a T-indistinguishability operator E on a finite set X when T is the minimum t-norm or an ordinal sum of a finite number of Łukasiewicz t-norms with covers the diagonal with as much precision as needed.

Indeed, we can consider the T-indistinguishability operator E' obtained in the last proposition as a T-indistinguishability operator valued on $L' = \{0, \frac{1}{n}, \frac{2}{n}, ..., \frac{n}{n} = 1\}$ so that we can calculate the dimension and a basis of E' (as a T-indistinguishability operator with T valued on L'). It is also a basis of E' as a T-indistinguishability operator with T valued on $[0, 1]$.

Example 13.43. Let E be the min-indistinguishability operator on $X = \{x_1, x_2, x_3, x_4\}$ with matrix

$$E = \begin{pmatrix} 1 & \frac{\sqrt{3}}{3} & 0 & 0 \\ \frac{\sqrt{3}}{3} & 1 & 0 & 0 \\ 0 & 0 & 1 & \frac{\sqrt{3}}{3} \\ 0 & 0 & \frac{\sqrt{3}}{3} & 1 \end{pmatrix}.$$

The matrix

$$E = \begin{pmatrix} 1 & \frac{1}{2} & 0 & 0 \\ \frac{1}{2} & 1 & 0 & 0 \\ 0 & 0 & 1 & \frac{1}{2} \\ 0 & 0 & \frac{1}{2} & 1 \end{pmatrix}$$

is a min-indistinguishability operator on X with $\|E - E'\| \leq \frac{1}{2}$. This E', considering the minimum t-norm on L, has been studied in Example 13.34. So E' as a T-indistinguishability operator valued on $[0, 1]$ has also dimension 2 and a basis of E' is $\{(\frac{1}{2}, 1, 0, 0), (0, 0, \frac{1}{2}, 1)\}$.

A

Appendix

Some Properties on t-Norms

This is a very short introduction to the basic properties of continuous t-norms. For more information about t-norms I suggest the book [83].

Definition A.1. *A continuous t-norm is a map* $T : [0,1] \times [0,1] \to [0,1]$ *such that for all* $x, y, z \in [0,1]$ *satisfies*

1. $T(T(x,y),z)) = T(x,T(y,z))$ *(Associativity)*
2. $T(x,y) = T(y,x)$ *(Commutativity)*
3. $T(1,x) = x$
4. T *is a non-decreasing map*
5. T *is a continuous map.*

NB. Commutativity can be derived from the other properties though the proof is not trivial.

Example A.2

1. The minimum t-norm min defined by $\min(x,y)$ for all $x, y \in [0,1]$.
2. The t-norm of Łukasiewicz defined by $T(x,y) = \max(0, x+y-1)$.
3. The Product t-norm $T(x,y) = x \cdot y$.

It is trivial to prove that the minimum t-norm is the greatest t-norm.

Definition A.3. *For a t-norm* T $x \in [0,1]$ *is an idempotent element if and only if* $T(x,x) = x$. $E(T)$ *will be the set of idempotent elements of* T.

Definition A.4. *A t-norm* T *is Archimedean if and only if* $E(T) = \{0,1\}$.

Example A.5. The t-norms Product and Łukasiewicz are Archimedean t-norms, while the minimum t-norm is not.

Definition A.6. *For a t-norm T, $x \in [0,1]$ is nilpotent if and only if there exists $n \in N$ such that $T^n(x) = 0$. $Nil(T)$ will be the set of nilpotent elements of T.*

(T^n is defined recursively: $T^n(x) = T(T^{n-1}(x), x)$).

Theorem A.7. *If a t-norm T is continuous Archimedean, then $Nil(T)$ is $[0,1)$ or $\{0\}$. In the first case, T is called non-strict Archimedean. In the second case, T is called strict Archimedean.*

Definition A.8. *Two t-norms T, T' are isomorphic if and only if there exists a bijective map $f : [0,1] \to [0,1]$ such that*

$$(f \circ T)(x,y) = T'(f(x), f(y))$$

Theorem A.9

- *All continuous strict Archimedean t-norms are isomorphic to the Product t-norm.*
- *All continuous non-strict Archimedean t-norms are isomorphic to the t-norm of Łukasiewicz.*

Theorem A.10. *Ling's Theorem*

A continuous t-norm T is Archimedean if and only if there exists a continuous and strictly decreasing function $t : [0,1] \to [0,\infty)$ with $t(1) = 0$ such that

$$T(x,y) = t^{[-1]}(t(x) + t(y))$$

where $t^{[-1]}$ is the pseudo inverse of t, defined by

$$t^{[-1]}(x) = \begin{cases} t^{-1}(x) & \text{if } x \in [0, t(0)] \\ 0 & \text{otherwise.} \end{cases}$$

T is strict if $t(0) = \infty$ and non-strict otherwise. t is called an additive generator of T and two generators of the same t-norm differ only by a positive multiplicative constant.

Example A.11

1. $t(x) = 1 - x$ is an additive generator of the t-norm of Łukasiewicz.
2. $t(x) = -\log(x)$ is an additive generator of the Product t-norm.

Theorem A.12. *Given a continuous t-norm T there exists a set of at most denombrable disjoint open intervals (a_i, b_i) such that in every set $[a_i, b_i] \times [a_i, b_i]$ the t-norm is a reduced copy T_i of an Archimedean t-norm and outside these sets the t-norm coincides with the minimum one. T is then called an ordinal sum of T_i.*

References

1. Aczél, J.: Lectures on functional equations and their applications. Academic Press, NY-London (1966)
2. Alsina, C., Castro, J.L., Trillas, E.: On the characterization of S and R implications. In: Proc. IFSA 1995 Conference, Sao Paulo, pp. 317–319 (1995)
3. Alsina, C., Trillas, E.: On natural metrics. Stochastica 2(3), 15–22 (1977)
4. Alsina, C., Trillas, E., Valverde, L.: On some logical connectives for fuzzy set theory. J. Math. Ann. Appl. 93, 15–26 (1983)
5. Bazaraa, M.S., Sherali, H.D., Shetty, C.M.: Nonlinear programming: theory and algorithms, 2nd edn. John Wiley & Sons, New York (1993)
6. Bellman, R.E., Kalaba, R., Zadeh, L.A.: Abstraction and pattern classification. J. Math. Anal. Appl. 13, 1–7 (1966)
7. Bělohlávek, R.: Fuzzy Relational Systems: Foundations and Principles. Kluwer Academic Publishers, New York (2002)
8. Bělohlávek, R.: Concept lattices and order in fuzzy logic. Annals of Pure and Applied Logic 128, 277–298 (2004)
9. Bělohlávek, R., Vychodil, V.: Algebras with fuzzy equalities. Fuzzy Sets and Systems 157, 161–201 (2006)
10. Bezdek, J.C., Harris, J.O.: Fuzzy partitions and relations: An axiomatic basis for clustering. Fuzzy Sets and Systems 1, 112–127 (1978)
11. Bhutani, K.R., Mordeson, J.N.: Similarity relations, vague groups, and fuzzy subgroups. New Math. Nat. Comput. 2, 195–208 (2006)
12. Bodenhofer, U.: A Similarity-Based Generalization of Fuzzy Orderings, Schriftenreihe der Johannes-Kepler-Universität Linz, vol. C26. Universitäts verlag, Rudolf Trauner (1999)
13. Boixader, D.: Contribució a l'estudi dels morfismes entre operadors d'indistingibilitat. Aplicació al raonament aproximat. Ph.D. Dissertation, UPC, In Catalan (1997)
14. Boixader, D.: T-indistinguishability operators and approximate reasoning via CRI. In: Dubois, D., Klement, E.P., Prade, H. (eds.) Fuzzy Sets, Logics and Reasoning about Knowledge, pp. 255–268. Kluwer Academic Publishers, New York (1999)
15. Boixader, D., Jacas, J.: Generators and Dual T-indistinguishabilities. In: Bouchon-Meunier, B., Yager, R.R., Zadeh, L.A. (eds.) Fuzzy Logic and Soft Computing, pp. 283–291. World Scientific, Singapore (1995)
16. Boixader, D., Jacas, J.: Extensionality based approximate reasoning. International Journal of Approximate Reasoning 19, 221–230 (1998)
17. Boixader, D., Jacas, J., Recasens, J.: Fuzzy equivalence relations: advanced material. In: Dubois, D., Prade, H. (eds.) Fundamentals of Fuzzy Sets, pp. 261–290. Kluwer Academic Publishers, New York (2000)

18. Boixader, D., Jacas, J., Recasens, J.: Upper and lower approximation of fuzzy sets. Int. J. of General Systems 29, 555–568 (2000)
19. Boixader, D., Jacas, J., Recasens, J.: Searching for meaning in defuzzification. Int. J. Uncertainty, Fuzziness and Knowledge-based Systems 7, 475–482 (1999)
20. Boixader, D., Jacas, J., Recasens, J.: A characterization of the columns of a T-indistinguishability operator. In: Proc. FUZZ IEEE 1998 Conference, Anchorage, pp. 808–812 (1998)
21. Bou, F., Esteva, F., Godo, L., Rodríguez, R.: On the Minimum Many-Valued Modal Logic over a Finite Residuated Lattice. Journal of Logic and computation (2010), doi:10.1093/logcom/exp062
22. Calvo, T., Kolesárova, A., Komorníková, M., Mesiar, R.: Aggregation Operators: Properties, Classes and Construction Methods. In: Mesiar, R., Calvo, T., Mayor, G. (eds.) Aggregation Operators: New Trends and Applications. Studies in Fuzziness and Soft Computing, pp. 3–104. Springer, Heidelberg (2002)
23. Castro, J.L., Klawonn, F.: Similarity in Fuzzy Reasoning. Mathware & Soft Computing 2, 197–228 (1996)
24. Castro, J.L., Trillas, E., Zurita, J.M.: Non-Monotonic Fuzzy Reasoning. Fuzzy Sets and Systems 94, 217–225 (1998)
25. Daňková, M.: Generalized extensionality of fuzzy relations. Fuzzy Sets and Systems 148, 291–304 (2004)
26. Dawyndt, P., De Meyer, H., De Baets, B.: The complete linkage clustering algorithm revisited. Soft Computing 9, 85–392 (2005)
27. De Baets, B., Kerre, E.: Fuzzy relations and applications. In: Hawkes, P. (ed.) Advances in Electronics and Electron Physics, vol. 89, pp. 255–324. Academic, New York (1994)
28. De Baets, B., Mareš, M., Mesiar, R.: T-partitions of the real line generated by idempotent shapes. Fuzzy Sets and Systems 91, 177–184 (1997)
29. De Baets, B., Mesiar, R.: T-partitions. Fuzzy Sets and Systems 97, 211–223 (1998)
30. De Baets, B., Mesiar, R.: Metrics and T-equalities. J. Math. Anal. Appl. 267, 531–547 (2002)
31. De Baets, B., De Meyer, H.: Transitive approximation of fuzzy relations by alternating closures and openings. Soft Computing 7, 210–219 (2003)
32. Delgado, M.: On a definition of fuzzy function. In: Proc. of First World Conference on Mathematics at the Service of Man, Barcelona, pp. 426–448 (1977)
33. Delgado, M.: Una nueva definición de aplicación difusa. Stochastica 4, 75–80 (1980) (in spanish)
34. Demirci, M.: Vague groups. J. Math. Anal. Appl. 230, 142–156 (1999)
35. Demirci, M.: Fuzzy functions and their fundamental properties. Fuzzy Sets and Systems 106, 239–246 (1999)
36. Demirci, M.: Fuzzy functions and their applications. J. Math. Anal. Appl. 252, 495–517 (2000)
37. Demirci, M.: Fundamentals of M-vague algebra and M-vague arithmetic operations. International Journal of Uncertainty, Fuzziness and Knowledge-Based Systems 10, 25–75 (2002)
38. Demirci, M.: Foundations of fuzzy functions and vague algebra based on many-valued equivalence relations part I: fuzzy functions and their applications, part II: vague algebraic notions, part III: constructions of vague algebraic notions and vague arithmetic operations. Int. J. General Systems 32, 123–155, 157–175, 177–201 (2003)

39. Demirci, M., Recasens, J.: Fuzzy Groups, Fuzzy Functions and Fuzzy Equivalence Relations. Fuzzy Sets and Systems 144, 441–458 (2004)
40. Di Nola, A., Sessa, S., Pedrycz, W., Sánchez, E.: Fuzzy relation equations and their applications to knowledge engineering. Kluwer Academic Publishers, New York (1989)
41. Dubois, D., Ostasiewicz, W., Prade, H.: Fuzzy Sets: History and basic notions. In: Dubois, D., Prade, H. (eds.) Fundamentals of Fuzzy Sets, pp. 21–124. Kluwer Academic Publishers, New York (2000)
42. Dubois, D., Prade, H.: Similarity Based Approximate Reasoning. In: Zurada, J.M., Marks II, R.J., Robinson, C.J. (eds.) Computational Intelligence Imitating Life, pp. 69–80. IEEE Press, Orlando (1994)
43. Erdös, P.: On the combinatorial problems which I would most like to see solved. Combinatoria 1, 25–42 (1981)
44. Esteva, F., Garcia, P., Godo, L.: Relating and extending semantical approaches to possibilistic reasoning. International Journal of Approximate Reasoning 10, 311–344 (1995)
45. Fodor, J.C., Roubens, M.: Structure of transitive valued binary relations. Math. Soc. Sci. 30, 71–94 (1995)
46. Foulloy, L., Benoit, E.: Building a class of fuzzy equivalence relations. Fuzzy Sets and Systems 157, 1417–1437 (2006)
47. Garmendia, L., Salvador, A., Montero, J.: Computing a T-transitive lower approximation or opening of a proximity relation. Fuzzy Sets and Systems 160, 2097–2105 (2009)
48. Garmendia, L., González del Campo, R., López, V., Recasens, J.: An Algorithm to Compute the Transitive Closure, a Transitive Approximation and a Transitive Opening of a Fuzzy Proximity. Mathware & Soft Computing 16, 175–191 (2009)
49. Godo, L., Rodriguez, R.O.: Logical approaches to fuzzy similarity-based reasoning: an overview. In: Riccia, G., Dubois, D., Lenz, H.J., Kruse, R. (eds.) Preferences and Similarities. CISM Courses and Lectures, vol. 504, pp. 75–128. Springer, Heidelberg (2008)
50. González del Campo, R., Garmendia, L., Recasens, J.: Estructuras de las similaridades. In: Proc. ESTYLF 2008, Langreo-Mieres, pp. 211–215 (2008) (in spanish)
51. González del Campo, R., Garmendia, L., Recasens, J.: Transitive closure of interval-valued relations. In: Proc. IFSA-EUSFLAT 2009, Lisboa, pp. 837–842 (2009)
52. Gottwald, S.: Fuzzy Sets and Fuzzy Logic: the Foundations of Application–from a Mathematical Point of Wiew. Friedr. Vieweg & Sohn Verlagsgesellschaft mbH, Wiesbaden (1993)
53. Gottwald, S., Bandemer, H.: Fuzzy Sets, Fuzzy Logic, Fuzzy Methods. Wiley, New York (1996)
54. Hajek, P.: Metamathematics of fuzzy logic. Kluwer Academic Publishers, New York (1998)
55. Hernández, E., Recasens, J.: A reformulation of entropy in the presence of indistinguishability operators. Fuzzy Sets and Systems 128, 185–196 (2002)
56. Hernández, E., Recasens, J.: Generating Indistinguishability Operators from Prototypes. Int J. Intelligent Systems 17, 1131–1142 (2002)

57. Hernández, E., Recasens, J.: A General Framework for Induction of Decision Trees under Uncertainty. In: Lawry, J., G. Shanahan, J., L. Ralescu, A. (eds.) Modelling with Words. LNCS (LNAI), vol. 2873, pp. 26–43. Springer, Heidelberg (2003)
58. Höhle, U.: The category CM-SET, an algebraic approach to uncertainty. Preprint, Cincinnati (1986)
59. Höhle, U.: Quotients with respect to similarity relations. Fuzzy Sets and Systems 27, 31–34 (1988)
60. Höhle, U.: M-valued sets and sheaves over integral commutative CL-monoids. In: Rodabaugh, S.E., Klement, E.P., Höhle, U. (eds.) Applications to Category Theory to Fuzzy Subsets, pp. 33–72. Kluwer Academic Publishers, Dordrecht (1992)
61. Jacas, J.: Contribucioó a l'estudi de les relacions d'indistingibilitat i a les seves aplicacions als processos de classificació. Ph.D. Dissertation, UPC (1987) (in catalan)
62. Jacas, J.: On the generators of T-indistinguishability operators. Stochastica 12, 49–63 (1988)
63. Jacas, J.: Similarity Relations - The Calculation of Minimal Generating Families. Fuzzy Sets and Systems 35, 151–162 (1990)
64. Jacas, J.: Fuzzy Topologies induced by S-metrics. The Journal of fuzzy mathematics 1, 173–191 (1993)
65. Jacas, J., Monreal, A., Recasens, J.: A model for CAGD using fuzzy logic. International Journal of Approximate Reasoning 16, 289–308 (1997)
66. Jacas, J., Recasens, J.: On some kinds of probabilistic relations. In: Bouchon-Meunier, B., Valverde, L., Yager, R.R. (eds.) Intelligent Systems with uncertainty, pp. 171–178. Elsevier, Amsterdam (1993)
67. Jacas, J., Recasens, J.: Fuzzy numbers and equality relations. In: Proc. FUZZ IEEE 1993, San Francisco, pp. 1298–1301 (1993)
68. Jacas, J., Recasens, J.: Fixed points and generators of fuzzy relations. J. Math. Anal. Appl. 186, 21–29 (1994)
69. Jacas, J., Recasens, J.: Fuzzy T-transitive relations: eigenvectors and generators. Fuzzy Sets and Systems 72, 147–154 (1995)
70. Jacas, J., Recasens, J.: One dimensional indistinguishability operators. Fuzzy Sets and Systems 109, 447–451 (2000)
71. Jacas, J., Recasens, J.: Maps and Isometries between indistinguishability operators. Soft Computing 6, 14–20 (2002)
72. Jacas, J., Recasens, J.: Aggregation of T-Transitive Relations. Int J. of Intelligent Systems 18, 1193–1214 (2003)
73. Jacas, J., Recasens, J.: The Group of Isometries of an Indistinguishability Operator. Fuzzy Sets and Systems 146, 27–41 (2004)
74. Jacas, J., Recasens, J.: A Calculation of Fuzzy Points. In: Proc. FUZZ'IEEE 2006, Vancouver (2006)
75. Jacas, J., Recasens, J.: Maps and isometries between indistinguishability operators. Soft Computing 6, 14–20 (2002)
76. Jacas, J., Recasens, J.: Aggregation Operators based on Indistinguishability Operators. Int. J. of Intelligent Systems 37, 857–873 (2006)
77. Jacas, J., Recasens, J.: E_T-Lipschitzian and E_T-kernel aggregation operators. Int. J. Approximate Reasoning 46, 511–524 (2007)
78. Jacas, J., Valverde, L.: On fuzzy relations, metrics and duster analysis. In: Verdegay, J.L., Delgado, M. (eds.) Approximate Reasoning Tools for Artificial Intelligence, pp. 21–38. ISR 96 Verlag TUV, Rheinland (1990)

79. Katz, M.: Inexact Geometry. Notre Dame J. Formal Logic 21, 521–535 (1980)
80. Klawonn, F.: Fuzzy Points, Fuzzy Relations and Fuzzy Functions. In: Nov'ak, V., Perfilieva, I. (eds.) Discovering the World with Fuzzy Logic, pp. 431–453. Physica-Verlag, Heidelberg (2000)
81. Klawonn, F., Kruse, R.: Equality Relations as a Basis for Fuzzy Control. Fuzzy Sets and Systems 54, 147–156 (1993)
82. Klawoon, F., Kruse, R.: Fuzzy control as interpolation on the basis of equality relations. In: Proc. of the FUZZ'IEE 1993, pp. 1125–1130 (1993)
83. Klement, E.P., Mesiar, R., Pap, E.: Triangular norms. Kluwer Academic Publishers, Dordrecht (2000)
84. Klir, G.J., Yuan, B.: Fuzzy sets and fuzzy logic: Theory and applications. Prentice Hall, Englewood Cliffs (1995)
85. Kruse, R., Gebhardt, J., Klawonn, F.: Foundations of Fuzzy Systems. Wiley, Chichester (1994)
86. Kumar De, S., Biswas, R., Ranjan Roy, A.: On extended fuzzy relational database model with proximity relations. Fuzzy Sets and Systems 117, 195–201 (2001)
87. Lawry, J.: Possibilistic Normalization and Reasoning under Partial Inconsistency. Int. J. Uncertainty, Fuzziness and Knowledge-based Systems 9, 413–436 (2001)
88. Lee, H.S.: An optimal algorithm for computing the max-min transitive closure of a fuzzy similarity matrix. Fuzzy Sets and Systems 123, 129–136 (2001)
89. Leibniz, G.W.: The PhilosophicalWorks of Leibnitz. Translated from the original Latin and French with notes by George Martin Duncan. Morehouse & Taylor, Publishers, Tuttle (1890)
90. Liu, Y., Kerre, E.: An overview of fuzzy quantifiers (I). Interpretations. Fuzzy Sets and Systems 95, 1–21 (1998)
91. Lowen, R.: Fuzzy topological spaces and fuzzy compactness. J. Math. Anal. Appl. 56, 623–633 (1976)
92. Mas, M., Monserrat, M., Torrens, J.: QL-Implications on a finite chain. In: Proc. Eusflat 2003, Zittau, pp. 281–284 (2003)
93. Mayor, G., Torrens, J.: Triangular norms on discrete settings. In: Klement, E.P., Mesiar, R. (eds.) Logical, Algebraic, Analytic, and Probabilistic Aspects of Triangular Norms, pp. 189–230. Elsevier, Amsterdam (2005)
94. Menger, K.: Untersuchungen über allgemeine Metrik. Math. Ann. 100, 75–113 (1928) (in German)
95. Menger, K.: Probabilistic theory of relations. Proc. Nat. Acad. Sci USA 37, 178–180 (1951)
96. Mesiar, R., Novák, V.: Operations fitting triangular-norm-based biresiduation. Fuzzy Sets and Systems 104, 77–84 (1999)
97. Mesiar, R., Reusch, B., Thiele, H.: Fuzzy equivalence relations and fuzzy partitions. J. Mult.-Valued Logic Soft Comput. 12, 167–181 (2006)
98. Miin-Shen, Y., Hsing-Mei, S.: Cluster analysis based on fuzzy relations. Fuzzy Sets and Systems 120, 197–212 (2001)
99. Moser, B.: On the T-transitivity of kernels. Fuzzy Sets and Systems 157, 1787–1796 (2006)
100. Moser, B., Winkler, R.: A relationship between equality relations and the T-redundancy of fuzzy partitions and its application to Sugeno controllers. Fuzzy Sets and Systems 114, 455–467 (2000)

101. Naessens, H., De Meyer, H., De Baets, B.: Algorithms for the Computation of T-Transitive Closures. IEEE Trans. Fuzzy Systems 10, 541–551 (2002)
102. Morsi, N.N., Yakout, M.M.: Axiomatics for fuzzy rough sets. Fuzzy Sets and Systems 100, 327–342 (1998)
103. Novák, V.: Fuzzy functions in fuzzy logic with fuzzy equality. Acta Math. Inform. Univ. Ostraviensis 9, 59–66 (2001)
104. Novák, V., Perfilieva, I., Močkoř, J.: Mathematical Principles of Fuzzy Logic. Kluwer Academic Publishers, Dordrecht (2000)
105. Ovchinnikov, S.: Representations of transitive fuzzy relations. In: Skala, H.J., Termini, S., Trillas, E. (eds.) Aspects of Vagueness, pp. 105–118. D. Reidel, Dordrecht (1984)
106. Ovchinnikov, S.: On the transitivity property. Fuzzy Sets and Systems 20, 241–243 (1986)
107. Ovchinnikov, S.: Similarity relations, fuzzy partitions, and fuzzy orderings. Fuzzy Sets and Systems 40, 107–126 (1991)
108. Ovchinnikov, S.: The duality principle in fuzzy set theory. Fuzzy Sets and Systems 42, 133–144 (1991)
109. Ovchinnikov, S.: Aggregating transitive fuzzy binary relations. Internat. J. Uncertain. Fuzziness Knowledge-Based Systems 3, 47–55 (1995)
110. Pawlak, Z.: Rough sets. Internat. J. Comput. Inform. Sci. 11, 341–356 (1982)
111. Perfilieva, I.: Fuzzy transforms: Theory and applications. Fuzzy Sets and Systems 157, 993–1023 (2006)
112. Poincaré, H.: La science et l'hypothèse. Flammarion, Paris (1902) (in French)
113. Pradera, A., Trillas, E., Castiñeira, E.: On the aggregation of some classes of fuzzy relations. In: Magdalena, L., Bouchon-Meunier, B., Gutiérrez-Ríos, J., Yager, R.R. (eds.) Technologies for Constructing Intelligent Systems 2: Tools. Studies in Fuzziness and Soft Computing, pp. 125–136. Springer, Heidelberg (2002)
114. Pradera, A., Trillas, E.: A note on Pseudometrics Aggregation. Int. J. General Systems 31, 41–51 (2002)
115. Prats, F., Rovira, R., Sánchez, M.: Indistinguishability operators with probabilistic values. Rev. Roumaine Math. Pures Appl. 35, 691–704 (1990)
116. Recasens, J.: On a geometric combinatorial problem. Discrete Mathematics 184, 273–279 (1998)
117. Recasens, J.: The Structure of Decomposable Indistinguishability Operators. Information Sciences 178, 4094–4104 (2008)
118. Recasens, J., Lawry, J.: Normalizing Possibility Distributions using t-norms. Int. J. Uncertainty, Fuzziness and Knowledge-based Systems 11, 343–360 (2003)
119. Riera, T.: How similarity matrices are? Stochastica 2, 77–80 (1978)
120. Rosenfeld, A.: Fuzzy Groups. J. Math. Anal. Appl. 35, 512–517 (1971)
121. Ruspini, E.: A new approach to clustering. Information and Control 15, 22–32 (1969)
122. Ruspini, E.: Recent development in fuzzy clustering. In: Yager, R.R. (ed.) Fuzzy Set and Possibility theory: Recent Developments, pp. 133–147. Pergamon Press, New York (1982)
123. Ruspini, E.: On the semantics of fuzzy logic. International Journal of Approximate Reasoning 5, 45–88 (1991)
124. Saaty, T.L.: The Analytic Hierarchy Process. McGraw-Hill, New York (1980)
125. Saminger, S., Mesiar, R., Bodenhofer, U.: Domination of aggregation operators and preservation of transitivity. International Journal of Uncertainty, Fuzziness and Knowledge-Based Systems 10, 11–35 (2002)

126. Solodovnikov, A.S.: Systems of Linear Inequalities. University of Chicago Press, Chicago (1980)
127. Schweizer, B., Sklar, A.: Probabilistic Metric Spaces. North-Holland, Amsterdam (1983)
128. Shenoi, S., Melton, A.: Proximity relations in the fuzzy relational database model. Fuzzy Sets and Systems 5, 31–46 (1981)
129. De Soto, A.R., Recasens, J.: Modelling a Linguistic Variable as a Hierarchical Family of Partitions induced by an Indistinguishability Operator. Fuzzy Sets and Systems 121, 57–67 (2001)
130. Tamura, S., Higuchi, S., Tanaka, K.: Pattern classification based on fuzzy relations. IEEE Trans. on Systems, Man and Cybernetics 1, 61–66 (1971)
131. Tan, S.K., Teh, H.H., Wang, P.Z.: Sequential representation of fuzzy similarity relations. Fuzzy Sets and Systems 67, 181–189 (1994)
132. Tomas, M.S., Alsina, C., Rubio-Martinez, J.: Pseudometrics from three-positive semidefinite similarities. Fuzzy Sets and Systems 157, 2347–2355 (2006)
133. Trillas, E., Alsina, C.: Introducción a los espacios métricos generalizados. Fund. March. Serie Universitaria 49 (Madrid 1978)
134. Trillas, E.: Sobre funciones de negación en la teoría de conjuntos borrosos. Stochastica 3, 47–59 (1979)
135. Trillas, E.: Assaig sobre les relacions d'indistingibilitat. Proc. Primer Congrés Català de Lògica Matemàtica, Barcelona, 51-59 (1982)
136. Trillas, E., Valverde, L.: An inquiry into indistinguishability operators. Theory Decis. Lib., pp. 231–256. Reidel, Dordrecht (1984)
137. Trillas, E., Valverde, L.: On Implication and Indistinguishability in the Setting of Fuzzy Logic. In: Kacprzyk, J., Yager, R.R. (eds.) Management Decision Support Systems Using Fuzzy Sets and Possibility Theory, pp. 198–212. Verlag TUV Rheinland Koln (1984)
138. Trillas, E., Valverde, L.: On modus ponens in fuzzy logic. In: Proc. of the 15th ISMVL Kingston, Ont., pp. 294–301 (1985)
139. Valverde, L.: On the Structure of F-indistinguishability Operators. Fuzzy Sets and Systems 17, 313–328 (1985)
140. Yager, R.R.: On ordered weighted averaging aggregation operators in multicriteria decision making. IEEE Trans Syst Man Cybern 18, 183–190 (1988)
141. Yager, R.R.: Entropy measures under similarity relations. Internat. J. General Systems 20, 341–358 (1992)
142. Yager, R.R., Filev, D.P.: Essentials of fuzzy modeling and control. John Wiley & Sons, Inc., New York (1994)
143. Zadeh, L.A.: Fuzzy sets. Information Control 8, 338–353 (1965)
144. Zadeh, L.A.: Similarity relations and fuzzy orderings. Inform. Sci. 3, 177–200 (1971)
145. Zadeh, L.A.: Fuzzy sets and information granularity. In: Gupta, M.M., Ragade, R.K., Yager, R.R. (eds.) Advances in Fuzzy Set Theory and Applications, pp. 3–18. North-Holland, Amsterdam (1979)
146. Zadeh, L.A.: Toward a theory of fuzzy information granulation and its centrality in human reasoning and fuzzy logic. Fuzzy Sets and Systems 90, 111–127 (1997)

Index